Helion & Company Limited
Unit 8 Amherst Business Centre
Budbrooke Road
Warwick
CV34 5WE
England
Tel. 01926 499 619
Email: info@helion.co.uk
Website: www.helion.co.uk
Twitter: @helionbooks
Visit our blog http://blog.helion.co.uk/

Text © Krzysztof Dabrowski 2023
Photographs © as individually credited
Colour artwork © Luca Canossa, Tom Cooper and Leon Manoucherians 2023
Maps © Tom Cooper 2023

Designed and typeset by Farr out Publications, Wokingham, Berkshire
Cover design Paul Hewitt, Battlefield Design (www.battlefield-design.co.uk)

Cover photo:: A Sukhoi Su-15TM interceptor of the PVO, armed with a pair each of R-98 (ASCC/NATO-codename 'AA-3 Anab') and R-60M ('AA-8 Aphid') air-to-air missiles, as seen while extending its airbrakes to slow down after overshooting a NATO reconnaissance aircraft. (Albert Grandolini Collection)

Cover artwork:: Weighing 43,000kg on take-off and 30m long, the Tupolev Tu-128 was a big, unsophisticated but fast, long-range and reliable interceptor, required to defend the 5,000-kilometres-'frontline' of the Soviet Siberia from attacks by US and British strategic bombers. It could reach Mach 1.5 and its standard armament consisted of four huge Bisnovat/Molniya R-4R/T (ASCC/NATO-codename 'AA-5 Ash') air-to-air missiles. This example served with the Amderma-based 72nd Guards Fighter Aviation Regiment from the late 1960s until the mid-1980s. (Artwork by Luca Canossa)

Every reasonable effort has been made to trace copyright holders and to obtain their permission for the use of copyright material. The author and publisher apologise for any errors or omissions in this work, and would be grateful if notified of any corrections that should be incorporated in future reprints or editions of this book.

ISBN 978-1-804510-27-8

British Library Cataloguing-in-Publication Data
A catalogue record for this book is available from the British Library

All rights reserved. No part of this publication may be reproduced, stored in a retrieval system, or transmitted, in any form, or by any means, electronic, mechanical, photocopying, recording or otherwise, without the express written consent of Helion & Company Limited.

We always welcome receiving book proposals from prospective authors.

CONTENTS

Abbreviations		2
Introductory Note		3
Acknowledgements		3
Introduction		3
1	Soviet Air Defence Force	4
2	Development of SAMs and ABM Defences	10
3	Fighter-Interceptor and Air-to-Air Weapons Development	14
4	Stratospheric Shenanigans	23
5	Borderline Behaviour	30
6	Balloon Busting	36
7	The Safest Mode of Transportation	39
8	Escape from Paradise	45
9	Destroy the Target at All Cost	47
10	On the Lighter Side	50
11	The Curtain Falls	52

Appendices
I	Commanders of the PVO since 1945	54
II	Aerial Victories of the PVO, 1945–1991	57

Documents	60
Bibliography	61
Notes	62
About the Author	66

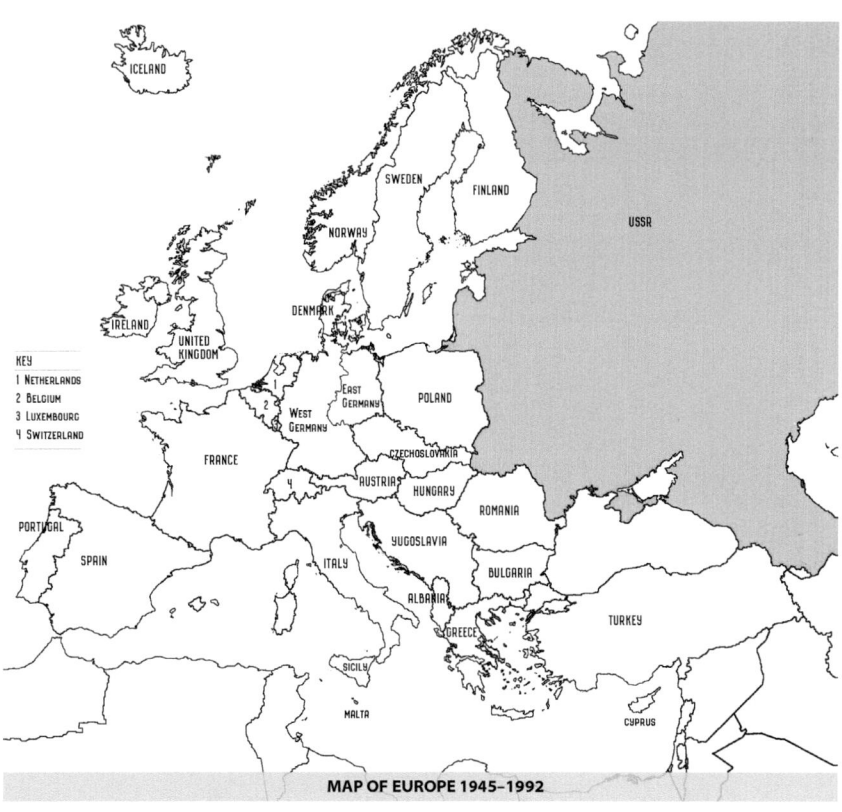

MAP OF EUROPE 1945–1992

Note: In order to simplify the use of this book, all names, locations and geographic designations are as provided in *The Times World Atlas*, or other traditionally accepted major sources of reference, as of the time of described events.

ABBREVIATIONS

AA	Anti-aircraft
AAM	Air-to-air missile
AB	Air Base
ABM	Anti-ballistic missile
AD	Air Defence
AF	Air Force
AFB	Air Force Base
ASCC	Air Standardisation Coordinating Committee
ATC	Air Traffic Control
ATMS	Automatic Tactical Management System
AWACS	Airborne Warning And Control System
BM	Ballistic Missile
BuNo	Bureau Number, a designation for aircraft assigned by the US Navy Bureau of Aeronautics
BVR	Beyond Visual Range
CAP	Combat Air Patrol
CIA	Central Intelligence Agency
C-in-C	Commander-in-Chief
c/n	Construction Number
CO	Commanding Officer
COMINT	Communications Intelligence
ECM	Electronic Counter Measures
ELINT	Electronic Intelligence
FSB	*Federalnaya Sluzhba Bezopasnosti* – Federal Security Service (the main security agency of the Russian Federation)
GCI	Ground Control Intercept
GIAP	*Gvardeyskiy Istrebitelniy Aviatsionniy Polk* – Guards Fighter Aviation Regiment (Soviet tactical fighter unit)
HUD	Head-up Display
HQ	Headquarters
IAP	*Istrebitelniy Aviatsionniy Polk* – Fighter Aviation Regiment (Soviet tactical fighter unit)
ICBM	Intercontinental Ballistic Missile
IFF	Identification Friend or Foe
Il	*Ilyushin* (the design bureau headed by Sergey Ilyushin)
IR	Infra-red
IRST	Infra-Red Search and Track (Infra-Red Sighting and Tracking) detecting and tracking of objects such as aircraft which give off infra-red radiation
JASDF	Japanese Air Self-Defence Force
JMSDF	Japanese Maritime Self-Defence
KGB	*Komitet Gosudarstvennoy Bezopasnosti* – State Security Committee (the main security agency of the Soviet Union)
MHz	Megahertz (10^6 Hz) with Hertz (Hz) being a unit of frequency defined as one cycle per second
MiG	Mikoyan i Gurevich (the design bureau led by Artyom Ivanovich Mikoyan and Mikhail Iosifovich Gurevich, also known as OKB-155 ormmZ 'Zenit')
MIRV	Multiple Independently Targetable Re-Entry Vehicle
NATO	North Atlantic Treaty Organization
PVO	*Protivovozdushnaya Oborona* literally air defence but frequently meant to denote the Air Defence Troops: see V-PVO
PRC	People's Republic of China, the Chinese communist state located on the Asian mainland
QRA	Quick Reaction Alert
RAF	Royal Air Force (of the United Kingdom)
RSW	Strategic Reconnaissance Wing
RWR	Radar Warning Receiver
SAC	Strategic Air Command
SAM	Surface-To-Air Missile
SAR	Search and Rescue
SAVAK	*Sazeman-e Eettelaat va Amniate-e Kesvar* is the Intelligence and Security Organization of the Country
Su	Sukhoi (the design bureau led by Pavel Ossipovich Sukhoi, also known as OKB-51)
TF	Task Force
TFW	Tactical Fighter Wing
THK	*Türk Hava Kuvvetleri* – Turkish Air Force
Tu	Tupolev (the design bureau headed by Andrei Tupolev, also known as OKB-156)
UAV	Unmanned Aerial Vehicle
USA	United States of America
USAF	United States Air Force
USN	United States Navy
USRJC	US– Russia Joint Commission on Pows/Mias
USSR	Union of Soviet Socialist Republics
UTC	Universal Time Coordinated
VHF	Very High Frequency that is from 30 to 300 mhz which corresponds with a wavelength from 1 to 10 m
V-PVO	*Voyska Protivovozdushnoy Oborony* also referred to as *Voyska Protivovozdushnoy Oborony Strany* – Soviet Air Defence Troops (Strany i.e. of the country) from 1954 in a separate branch of the Soviet military tasked with protecting the country's airspace
VVS	*Voyenno-Vozdushnyye Sily* – the Soviet Air Force
Yak	Yakovlev (the design bureau led by Alexander Yakovlev, also known as OKB-115 ormmZ 'Skorost')

INTRODUCTORY NOTE

The first volume's narrative closed with the description of incidents which had taken place in 1960. Likewise, the description of the Soviet air defence organisational development and arsenal also finished at that point in time i.e., the year 1960. The present volume picks up right where the other ended, with the caveat that some overlap is unavoidable. It aims to provide an account of Soviet air defences' structural evolution, changes in its weaponry as well as technology. Last but not least, it describes aerial incidents taking place right up to the demise of the Soviet Union. It should be noted that limits to matters treated herein, have been consciously set and are related to the subject at hand, geography and the temporal dimension.

First of all, other branches of the Soviet military are purposefully omitted with only the Soviet Air Defence Force (PVO) being focused on. In addition, solely incidents or similar events which took place, either in Soviet airspace or in its immediate vicinity, will be described. Likewise, the deployment and actions of Soviet PVO personnel in Cuba, Vietnam, Egypt, Syria, Libya, and other foreign countries shall also be omitted. Finally, regarding temporal limits, the ending coincides with the dissolution of the Soviet Union on 26 December 1991. That said, a few exceptions to these guiding principles were made when the matters described warranted as such.

ACKNOWLEDGEMENTS

It is impossible for a project such as this to materialise in the tangible form of a book without many helping hands landing their support, advice, and practical assistance. Thus, the author finds it indispensable to express his most sincere gratitude to Tom Cooper, Dymitry Zubkov, Robert S. Hopkins III, Jaroslaw Malinowski, Leon Manoucherians (editor of the Iranian Aviation Review), Arda Mevlutoglu, Adrian Symonds, Leonardo Daniel Nemec, as well as many others, thanks to whom this volume came into being.

INTRODUCTION

The shooting-down of F. G. Powers' U-2 in the vicinity of Sverdlovsk (modern Yekaterinburg) on 1 May 1960, was a dramatic climax of high-altitude aerial reconnaissance missions over the Soviet Union. However, the Cold War did not end with this and consequently, neither did its aerial dimension. On the contrary, the Cold War would continue for the next three decades sometimes escalating uncomfortably almost into a 'hot' war while at other times, relaxing into detente. With it, American and other Western aerial reconnaissance activity directed at the USSR, also continued, gaining or reducing in intensity yet never ceasing entirely. Obviously, this was not an end unto itself but served as a defensive purpose to avert any Soviet surprises. It was also an offensive one in preparation for an attack (even if one was never carried out) against the USSR; this being so until the Soviet state ceased to be.

During the period in question, the ballistic missile became the primary weapon of strategic nuclear attack. Measures to detect such and provide limited defences against them, will also be described herein. However, manned aviation was still a potent threat and combating it remained the PVO's primary task. The challenge it posed evolved due to a number of developments. One was the introduction of air launched cruise missiles which gave the strategic bomber new capabilities enabling it to strike at multiple targets from a stand-off range. Moreover, cruise missiles could also be launched from other platforms.

In addition, once fielded, guided munitions significantly changed the conduct of aerial bombardment. If in earlier times, the destruction of a target required either a nuclear bomb or a large number of aircraft dropping conventional ones, guided munitions made it possible to effectively incapacitate a target by hitting it with precision, where it mattered most, by a relatively small number of aircraft. Low level, deep penetration strikes by aircraft such as the F-111 presented yet another serious challenge. The threat of air attack upon the Soviet Union, whilst a grave concern, remained theoretical for it would have only materialised in case of a Third World War. However, we were all spared, what would have likely been, nothing short of Doomsday.

Nevertheless, aerial reconnaissance along USSR's periphery was an almost daily occurrence and it had to be countered. In addition, very capable reconnaissance satellites could not provide certain types of data. For an example, manned aircraft were best suited to register radar emissions and gather other information related to how the Soviet military, the PVO in particular, operated. In order to obtain such data, reconnaissance aircraft sometimes had to conduct their missions in a deliberately provocative way in order to goad the Soviet air defence apparatus into action. Needless to say, such undertakings carried a great risk for the aircraft and the aircrews involved.

Other forms of reconnaissance, such as photographic, were not entirely abandoned either. The aircraft performing these missions differed greatly ranging from high-altitude, Lockheed SR-71A Blackbird three-sonic strategic reconnaissance aircraft to low-flying, tactical fighter-reconnaissance types and large, lumbering propeller-driven reconnaissance aircraft were also frequently utilised. Apart from military aircraft there were also civilian airspace violators as well as aerostats which floated into Soviet airspace, and such also had to be dealt with.

As outlined, the Soviets were faced with numerous potential wartime threats as well as, ever-present peacetime challenges. By 1960, they had organised a national air defence force and were providing it with an ever-growing number of radars, surface-to-air missiles and manned fighter-interceptors. However, time did not stand still: the threats and challenges evolved, and the Soviet air defence arsenal had to evolve with them. Newer air defence weapons were continuously being developed and fielded which from time to time, were called into action when a given situation escalated. The development of Soviet air defences as well as incidents – sometimes tragic ones – when they were actually used in anger, are related herein.

1
SOVIET AIR DEFENCE FORCE

The *Voyska Protivovozdushnoy Oborony* (V-PVO, colloquially 'PVO') – the Soviet Air Defence Force – was, since its inception, functionally composed of three principal elements: *Istrebitelnaya Aviatsiya* (Fighter Aviation), *Zenitnyye Raketnyye Voyska* (Anti-Aircraft Missile Force, or missile troops) and *Radiotekhnicheskiye Voyska* (Radio-Technical or radar troops). Thus, the V-PVO had all the assets needed for the task of defending the airspace of the Soviet Union: it could detect threats and engage them either by means of SAMs or with fighter aircraft. Its responsibilities were also clearly delimited: ground forces units and fleets were left only with organic AD capabilities for their own protection from air attack but not that of the country's territory, similarly fighter units of the VVS (Air Force) had a tactical role separate from that of the PVO fighter aviation's sole mission of defending Soviet air space. Organisationally, the PVO had two Air Defence Districts as well as a number of Separate (Independent) Air Defence Armies.[1] Due to the threat posed by ballistic missiles and the resulting development of ABM defences, a fourth V-PVO branch the *Voyska protivoraketnoy i protivokosmicheskoy oborony* (anti-missile and space defence troops) was officially established in 1967.[2] The PVO's principal structure of AD districts and AD armies tasked with the defence of USSR's airspace, was maintained till 1979. Whilst not perfect, the PVO's structural and organisational arrangements proved to be sound and served their main purpose reasonably well. As an old adage says *'if ain't broke, don't fix it'* this being proven right by experience time and again. Yet the Soviets decided to go against this old wisdom and found out the hard way, that it was considered a wisdom for a reason.

For example, in 1978 the Chief of the General Staff and First Deputy Minister of Defence Marshal of the Soviet Union, Nikolai Ogarkov concocted the idea of a comprehensive reorganisation of the PVO, this being carried out in the years 1979–81.[3] In course of the said reorganisation, the USSR's territory was, for the purpose of air defence, divided into border (frontier) and internal areas, the Baku Air Defence District was disbanded as were a number of AD armies. This was accompanied by a great reshuffling of assets, in particular, numerous PVO fighter units were transferred to the VVS. Such a reorganisation of a more or less, well-functioning military apparatus brought about its disorganisation and as a consequence, the efficiency of Soviet air defences suffered greatly.

Soon, this lamentable state of affairs became obvious, resulting in the calls to do something about it, growing louder and louder. Thus in 1986, the previous structure of the PVO was in essence, re-established, though with a number of changes.[4] By the late 1980s, the Soviet Air Defence Force had the following basic organisational structure:

PVO HQ Moscow
- Moscow Air Defence District (Moscow)
- 2nd Independent Air Defence Army (HQ Minsk)
- 3rd Independent Special Purpose Rocket (Missile) Attack Warning Army (HQ Solnechnogorsk)
- 4th Independent Air Defence Army (HQ Sverdlovsk; modern-day Yekaterinburg)
- 6th Independent Air Defence Army (HQ Leningrad; modern-day St. Petersburg)
- 8th Independent Air Defence Army (HQ Kiev)
- 9th Independent Anti-Rocket (Missile) Defence Corps (HQ Akulovo)
- 10th Independent Air Defence Army (HQ Arkhangelsk)
- 11th Independent Air Defence Army (HQ Khabarovsk)
- 12th Independent Air Defence Army (HQ Tashkent)
- 14th Independent Air Defence Army (HQ Novosibirsk)
- 18th Independent Space Control Corps (HQ Noginsk)
- 19th Independent Air Defence Army (HQ Tbilisi; former Baku Red Banner Air Defence District)

Generally, an AD army would have a number of subordinate AD corps, which in turn, were composed of AD divisions and other units of lower order, though in some cases, AD divisions were under the direct control of a given AD army. The Moscow AD District had in its structure, an AD army solely tasked with defending the Soviet capital which operated the S-25 SAM rings around Moscow, until they were withdrawn from service and replaced by S-300P SAMs (ASCC/NATO-codename 'SA-10 Grumble'; see the following for details). In addition, ballistic missile attack warning ABM defences, as well as space surveillance assets, were grouped together in one separate (independent) army and two corps.

Airspace Surveillance, Intercept-Control and Strategic Early Warning
In order to fulfil their primary mission of detecting aerial threats, the PVO's radar units had at their disposal, a broad range of VHF air surveillance radars including the P-12 and P-18 (both received the ASCC/NATO-codename 'Spoon Rest'), as well as the long-ranged P-14 (ASCC/NATO-codename 'Tall King'). Towards the end of 1960s, the P-70 long-range surveillance radar with a detection range of 2,300km, entered service and it was, by the Soviets, regarded as one of their best types in this category.[5] Later, during the 1980s, the highly capable *Nebo*-series (ASCC/NATO-codename 'Box Spring') radars were fielded. There was a plethora of other radar types in the PVO's inventory: for example, the PRV-11 (ASCC/NATO-codename 'Side Net') and PRV-13 (ASCC/NATO-codename 'Odd Pair') were height-finders supplementing surveillance, ground control intercept and SAM radars, to provide a three-dimensional radar picture.

As a result of both quantitative and qualitative radar development, the Soviets were eventually able to set-up enough radar sites to achieve radar coverage along their entire 60,000km (37,000 miles) long border. However, for a number of reasons it was never perfect: the earth curvature and various topographic features limited or locally obstructed the coverage with various other factors also negatively impacting it on different occasions. In addition to the types mentioned, an important part in providing an effective air defence, in particular when utilising fighter-interceptors for that task, was also played by radars combining early warning/surveillance and ground control intercept functions, of which the P-35 and P-37 (both having the ASCC/NATO-codename 'Bar Lock') were the most numerous ones. The aforementioned deployment of interceptors and also of other assets such as surface-to-air missiles, required command and control systems which would tie them with radars into one functional-wise entity.

First to enter service was the *Vozdukh* system which was continuously modernised.[6] Thanks to subsequent improvements in technology of the new system: the *Luch* as well as *Rubezh* were also fielded. Those systems facilitated the best use of AD assets available, in particular through feeding via a radio command link, data provided by radars directly into fighter aircraft's on-board systems that enabled them to engage a target in the optimal position. For this purpose, Soviet interceptors serving with PVO fighter units, were appropriately equipped to work with the aforementioned systems. In the course of the systems continuous development, a high degree of automation was achieved, eventually allowing an interceptor to be literally, remotely flown from the ground.

However, as with radar coverage, the use of the systems in practice suffered from various limitations. In addition, pilots generally disliked fully automated guidance and as a rule, preferred to handle their aircraft performing whatever tasks were necessary, themselves receiving instructions by way of voice commands from ground controllers. The quality of guidance provided by the latter differed considerably depending on their skill, experience, as well as the hard-to-define ability to judge a situation and give the right commands at the right time. Putting it colloquially, some ground controllers 'had it' whilst others did not.[7] Yet even with all the caveats and limitations, the ground-based radar network, as well as associated command and control systems, provided the Soviet air defences with noteworthy capabilities.

In addition to an extensive network of radar sites, the Soviets also developed early warning and control aircraft for the PVO starting with the Tupolev Tu-126 (ASCC/NATO-codename 'Moss').[8] The 'red AWACS' utilised the airframe of a Tu-114 (ASCC/NATO-codename 'Cleat') passenger aircraft powered by four turboprop engines which had counter-rotating propellers and for the new role, was also fitted with an in-flight refuelling probe. In order to fulfil its primary task, the Tu-126 was equipped with the *Lyana* (ASCC/NATO-codename 'Flat Jack') radar system with a rotodome antenna mounted atop the fuselage, this being a typical configuration for such aircraft. Large targets could be detected from a distance of ca 300-400km, but the radar had problems with making out small ones especially against ground clutter.

A Tu-126's crew compromised six airmen as well as the same number of radar and electronic systems operators. Their working

The core of a typical ground control station of the PVO in the 1980s: notable are multiple displays, each showing the picture from a different radar system. This was necessary because the number of radars capable of presenting a three-dimensional image on their displays was minimal. Moreover, the PVO intentionally sought to have redundancy in case one radar would be destroyed in combat or disrupted by electronic countermeasures. The mass of ground controllers guiding interceptors were highly-qualified, mid-ranking officers and pilots were indoctrinated to follow their orders to the last dot and comma. (Albert Grandolini Collection)

An extremely rare photograph showing the inside of a Tu-126 airborne early warning aircraft, with an operator and one of the radar consoles. Whilst certainly a major enhancement for the PVO, the type was reportedly, tough on its crew. Noise levels were very high and radar operators had to serve long shifts with no sleeping bunks and barely adequate rest rooms. (Albert Grandolini Collection)

conditions were poor due to insufficient isolation from noise, vibrations, and radiation. Overall, nine Tu-126s were manufactured including a single prototype. They were grouped together in an independent (separate) squadron, composed of two flights each numbering four aircraft with the prototype acting as a backup to be used if an extra one was required. The unit was based at Šiauliai AB in the Soviet Lithuania but had regular detachments at Olenegorsk AB for exercises with PVO's fighter-interceptors based in the Far North. Aside those yearly deployments, the Tu-126s did not fly regular missions but instead sortied as required to support Soviet aerial activity or to monitor that of NATO.

The Tu-126s served for almost 20 years from 1965 till 1984. They were finally retired with the caveat that the prototype was pulled from service earlier to be used in the development work of

The Tupolev Tu-126 was the result of the first successful Soviet effort to develop an airborne early warning aircraft. In comparison to contemporary Western-made AWACS, its detection performance was poor: nevertheless, it was offering a large improvement in performance of the PVO – especially when operated in support of Tu-128s. (Albert Grandolini Collection)

a new Soviet early warning and control aircraft. Since deficiencies of the Tu-126 were obvious, a better aircraft and radar system of this kind were clearly needed. Work on this was entrusted to the Beriev design bureau which chose the Ilyushin Il-76 (ASCC/NATO-codename 'Candid') heavy transport as a platform for the project.

The new early warning and control aircraft, Beriev A-50 (ASCC/NATO-codename 'Mainstay') was fitted with a comprehensive suite of electronics, the centrepiece of which was the *Shmel* (ASCC/NATO-codename 'Squash Dome') radar system. It has a prominent rotodome antenna, can detect large targets from up to 650km, small ones such as for example cruise missiles from ca 200km. Moreover, in addition to a greater detection range in comparison to its predecessor, it can also deal much better with ground clutter. An A-50 is crewed by 15 men including 10 radar and electronic systems operators. The aircraft and its systems development were protracted, starting in the 1970s with the first units becoming operational in 1984. However, it was only officially accepted into service in 1989. Currently, the A-50 still serves with the Russian Air and Space Force and it was also used as the basis for the development of a new Russian early warning and control aircraft, Beriev A-100.

Apart from the threat posed by bombers and other manned strike aircraft, the Soviets were also faced with another very serious challenge, namely that of ballistic missiles. In the post-war years, the ballistic missile was 'mated' with nuclear warheads as soon as compact ones became available. Thus, in the 1950s the United States fielded a number of ballistic missiles with a nuclear war load: PGM-11 Redstone short-range ballistic missiles and PGM-19 Jupiter medium-range ballistic missiles. These were soon followed by intercontinental ballistic missiles such as the SM-65 Atlas as well

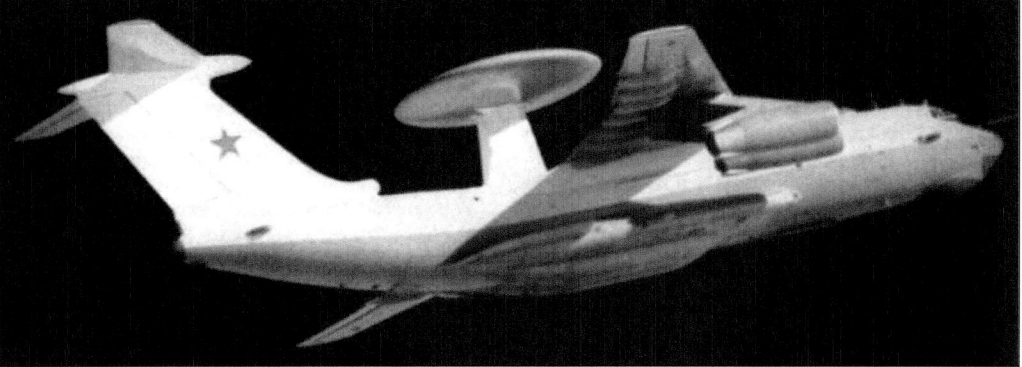

One of first clear photographs of an A-50 AWACS aircraft, taken in the mid-1980s over the Baltic Sea. In the West, the jet was promptly assessed as comparable in capabilities to the Boeing E-3 Sentry and declared as substantially improving the Soviet battle management capability – especially in regards of providing early warning against low-altitude penetration. Actually, the Soviets foremost used it as a type of air traffic control. (US DoD)

as later types. If that was not enough, starting with the UGM-27 Polaris, submarine launched ballistic missiles were also added to the West's nuclear arsenal. For the Soviets, it became imperative to be able to detect an incoming ballistic missile attack. A decision on the development of appropriate means to that end, was made in 1956.[9] However, it took some time before tangible results were achieved.

The first two ballistic missile early warning radar systems were built between 1965-69 in the vicinity of Olenegorks on the Kola Peninsula and Skrunda in Latvia (a Soviet Republic in those times). Both radars were of the *Dnestr-M* (ASCC/NATO-codename 'Hen House') type, each had two 244m long and 20m high arrays, with a building in the middle which housed computers, operators' stations et cetera When seen from above, the whole structure resembled a boomerang. The radars had a range of ca 2500-3000km, could provide warning of incoming ballistic missiles including ones launched from submarines in the Norwegian Sea as well as North Sea and became operational in 1970.[10]

As a next step, the Soviets built four *Dnepr* (which also had the ASCC/NATO-codename designation 'Hen House') early warning radars that were of a similar configuration as the earlier *Dnestr-M*. These radar facilities were located at: Balkhash in Kazakhstan (formerly a Soviet Republic), Mishelevka in Siberia (a space surveillance radar was also located there), Skrunda (in addition to the *Dnestr-M* system) and Sevastopol in Crimea. One more Dnepr radar site was subsequently built at Mukachevo in the Ukraine (a Soviet Republic back than), which together with the other *Dneprs*, *Dnestr-Ms* and Moscow's ABM radars (see below), by 1979 formed the USSR's ballistic missile attack early warning system.

In the 1980s, new ballistic missile early warning radar systems were built in the vicinity of Pechora (the town) and Gabala in Azerbaijan (former Soviet Republic). Each of the radar systems were designated *Daryal* – interestingly, and breaking with the usual naming practice, its western designation was Pechora after one of its sites – had two separate arrays: one measuring 100m by 100m and the other 40m by 40m. As many as eight such radar sites were planned, with some being under construction at the time of USSR's demise, but only the two aforementioned, *Daryal* radar facilities, were completed.

The ground-based radar stations described were augmented by early warning satellites, the deployment of which started in the late 1970s. Finally, the Soviets also started the construction of an optical space surveillance system meant to detect and track satellites or other objects in space by means of powerful telescopes mounted in domes. Interestingly, some Western analysts initially opined that these were high-energy laser anti-satellite weapons. As stated, the system known as *Okno*, was being developed in the Soviet Union but only became operational after its demise.[11]

COMPOSITION OF AN AIR DEFENCE ARMY

Perhaps the best-known army-level unit of the PVO was the 10th Independent Air Defence Army, responsible for the protection of the Kola Peninsula, the Archangelsk and the Severomorsk areas, all of which contained a massive concentration of Soviet naval facilities, including some of most important shipyards and military bases. As of 1968, the 10th Independent Air Defence Army was organised as depicted in Table 1.[12]

Table 1: Order of Battle of the 10th Independent Air Defence Army, PVO, 1968		
Unit	Element	Notes
21st Air Defence Corps		HQ Severomorsk
	941st Fighter Aviation Regiment	Kilpyavr AB, 3 squadrons of Su-9
	174th Guards Fighter Aviation Regiment	Monchegorsk AB; 3 squadrons of MiG-19 and Yak-28P
	431st Fighter Aviation Regiment	Afrikanda AB, 3 squadrons of MiG-19
	5 anti-aircraft missile regiments	
4th Air Defence Division		HQ Belushya Guba
	911th Fighter Aviation Regiment	Rogachevo AB (Novaya Zemlya), 3 squadrons of MiG-17
	72nd Guards Fighter Aviation Regiment	Amderma AB, 3 squadrons of MiG-19 and Yak-25M
	1 anti-aircraft missile regiment	
5th Air Defence Division		HQ Petrozavodsk
	265th Fighter Aviation Regiment	Poduzhemye AB, 3 squadrons of MiG-17
	641st Guards Fighter Aviation Regiment	Besovets AB (Petrozavodsk), 2 squadrons of MiG-17 and Yak-28P
	1 anti-aircraft missile regiment	
23rd Air Defence Division		HQ Arkhangelsk
	518th Fighter Aviation Regiment	Talagi AB, 3 squadrons of Tu-128
	524th Fighter Aviation Regiment	Letnoezersky/Obozersk AB, 3 squadrons of MiG-17 and Yak-28P
	3 anti-aircraft missile regiments	

Over the following years, the 10th Independent Air Defence Army was largely re-equipped, foremost with manned long-range interceptors and also by ground-based, long-range air defence systems like S-200 (ASCC/NATO-codename 'SA-5 Gammon'). Of particular importance were Tupolev Tu-128 supersonic interceptors (ASCC/NATO-codename 'Fiddler'), each of which was armed with two R-4R semi-active, radar homing and two R-4T infra-red homing, air-to-air missiles. Crucial for effectiveness of Tu-128s, was the deployment of the Vozdukh-1M automatic tactical management system (ATMS).

Originally developed to support operations of the Sukhoi Su-9 interceptors, at the core of the Vozdukh-1M were two computers: one used to track airborne targets, the other to compute interceptor vectors. Using the Lasour data link protocol, Vozdukh-1M could simultaneously track up to 82 targets and process all the available data to enable the computation of an optimal flight path for manned interceptors. A single Vozdukh-1M of the 10th Independent Air Defence Army proved capable of networking several Tu-128s, operating at 900–1000km away from their bases – and thus, well outside the range of ground-based radars or other guidance from the ground – into effective intercepts of incoming bombers. The system enabled similar operations of Sukhoi Su-15 interceptors, albeit at a much shorter range from their bases. By 1973, the 10th Army was thus composed as described in Table 2.

Continued on page 8

Table 2: Order of Battle of the 10th Independent Air Defence Army, PVO, 1973		
Unit	Element	Notes
21st Air Defence Corps		HQ Severomorsk
	941st Fighter Aviation Regiment	Kilpyavr AB, 3 squadrons of Su-9
	174th Guards Fighter Aviation Regiment	Monchegorsk AB; 3 squadrons of MiG-19 and Yak-28P
	431st Fighter Aviation Regiment	Afrikanda AB, 3 squadrons of MiG-19PM
	5 anti-aircraft missile regiments	
4th Air Defence Division		HQ Belushya Guba
	911th Fighter Aviation Regiment	Rogachevo AB (Novaya Zemlya), 3 squadrons of MiG-17
	72nd Guards Fighter Aviation Regiment	Amderma AB, 3 squadrons of Tu-128
	1 anti-aircraft missile regiment	
5th Air Defence Division		HQ Petrozavodsk
	265th Fighter Aviation Regiment	Poduzhemye AB, 3 squadrons of Su-15
	641st Guards Fighter Aviation Regiment	Besovets AB (Petrozavodsk), 2 squadrons of MiG-17 and Yak-28P
	1 anti-aircraft missile regiment	
23rd Air Defence Division		HQ Arkhangelsk
	518th Fighter Aviation Regiment	Talagi AB, 3 squadrons of Tu-128
	524th Fighter Aviation Regiment	Letnoezersky/Obozersk AB, 3 squadrons of MiG-17 and Yak-28P
	3 anti-aircraft missile regiments	

The growth of the 10th Independent Air Defence Army was further intensified during the second half of the 1970s, when all of its units had been re-equipped with radar-equipped interceptors – including first Mikoyan i Gurevich MiG-23Ms. These were the first jets in Soviet service whose radar and fire-control system had the look-down/shoot-down capability enabling them to target low-flying objects. In turn, the effectiveness of the MiG-23M was heavily dependent on support of such ATMS like the Vozdukh-1M. MiG-23Ms were closely followed by MiG-25P/PDS: the latter were considered of crucial importance because of operations of SR-71 Blackbird reconnaissance aircraft over the Barents Sea. Correspondingly, by 1983, the 10th Army was re-equipped as follows:

Table 3: Order of Battle of the 10th Independent Air Defence Army, PVO, 1983		
Unit	Element	Notes
21st Air Defence Corps		HQ Severomorsk
	941st Fighter Aviation Regiment	Kilpyavr AB, 3 squadrons of MiG-23M
	174th Guards Fighter Aviation Regiment	Monchegorsk AB; 3 squadrons of Yak-28P; in the process of conversion to the MiG-31
	431st Fighter Aviation Regiment	Afrikanda AB, 3 squadrons of Su-15TM
	5 anti-aircraft missile regiments	
4th Air Defence Division		HQ Belushya Guba
	911th Fighter Aviation Regiment	Rogachevo AB (Novaya Zemlya), 3 squadrons of Yak-28P
	72nd Guards Fighter Aviation Regiment	Amderma AB, 3 squadrons of Tu-128

Table 3: Order of Battle of the 10th Independent Air Defence Army, PVO, 1983 (*continued*)		
	1 anti-aircraft missile regiment	
5th Air Defence Division		HQ Petrozavodsk
	265th Fighter Aviation Regiment	Poduzhemye AB, 3 squadrons of Su-15TM
	641st Guards Fighter Aviation Regiment	Besovets AB (Petrozavodsk), 2 squadrons of Su-15TM
	1 anti-aircraft missile regiment	
23rd Air Defence Division		HQ Arkhangelsk
	518th Fighter Aviation Regiment	Talagi AB, 3 squadrons of Tu-128
	524th Fighter Aviation Regiment	Letnoezersky/Obozersk AB, 3 squadrons of MiG-25PDS
	3 anti-aircraft missile regiments	

Although Moscow found it ever harder to replace older interceptor types on one-for-one basis with new-, advanced-, but also more expensive aircraft, during the mid-1980s, the 10th Independent Air Defence Army was re-equipped once again. This time it was with an entirely new generation of fighter jets, all of which were equipped with long-range pulse-Doppler radars that had the look-down/shoot-down mode and missiles with the range of more than 50km. Correspondingly, by 1988, the composition of the 10th Army was as listed in Table 4.

Table 4: Order of Battle of the 10th Independent Air Defence Army, PVO, 1988		
Unit	Element	Notes
21st Air Defence Corps		HQ Severomorsk
	941st Fighter Aviation Regiment	Kilpyavr AB, 3 squadrons of Su-27
	174th Guards Fighter Aviation Regiment	Monchegorsk AB; 3 squadrons of MiG-31
	431st Fighter Aviation Regiment	Afrikanda AB, 3 squadrons of Su-15TM
	5 anti-aircraft missile regiments	
4th Air Defence Division		HQ Belushya Guba
	641st Guards Fighter Aviation Regiment	Rogachevo AB (Novaya Zemlya), 1 squadron of Su-27, 2 squadrons of Yak-28P
	72nd Guards Fighter Aviation Regiment	Amderma AB, 1 squadron of MiG-31B, 2 squadrons of Tu-128
	1 anti-aircraft missile regiment	
5th Air Defence Division		HQ Petrozavodsk
	265th Fighter Aviation Regiment	Poduzhemye AB, 3 squadrons of Su-15TM
	641st Guards Fighter Aviation Regiment	Besovets AB (Petrozavodsk), 3 squadrons of Su-15TM
	1 anti-aircraft missile regiment	
23rd Air Defence Division		HQ Arkhangelsk
	518th Fighter Aviation Regiment	Talagi AB, 3 squadrons of MiG-31B
	524th Fighter Aviation Regiment	Letnoezersky/Obozersk AB, 3 squadrons of MiG-25PDS
	3 anti-aircraft missile regiments	

2
DEVELOPMENT OF SAMs AND ABM DEFENCES

Soviet SAMs were deployed to protect specific locations such as large population centres, industrial facilities and major military bases. Whenever necessary, units operating SAMs could also be deployed to provide overlapping coverage of selected regions of particular interest. The former was prevalent in the early years of the PVO's anti-aircraft rocket (missile) troops, when the number of available SAMs was still very limited. Over the passage of time, both the number of SAM batteries and SAM types, especially long-range ones (see next paragraph) grew, it became possible to attain the latter. In particular 'belts' and 'clusters' of SAM sites provided area coverage for most regions of the USSR's European part, with SAM 'spots' in the Far North, the Caucasus, Central Asia and Siberia, Far East as well as other regions deemed important enough to warrant such protection. In contrast, ABM defences were only deployed around Moscow due to the *Anti-Ballistic Missile Treaty*[1].

PVO's Surface-to-Air Missile Arsenal
By 1960, the PVO had two operational SAM systems: the S-25 (ASCC/NATO-codename 'SA-1 Guild') and the S-75 (ASCC/NATO-codename 'SA-2 Guideline'). Whilst the former was solely tasked with the defence of Moscow being deployed on fixed sites ringing the Soviet capital, the latter formed the backbone of USSR's SAM defences across the country.[2] The S-75 proved its worth by among others, shooting-down a U-2 high-altitude reconnaissance aircraft on 1 May 1960 in course of a well-known Cold War incident, However, it had serious shortcomings; namely, it was unable to effectively engage low-flying targets and had a limited range. In particular, the former was judged to be a critical deficiency. For this reason, the S-125 *Neva* SAM system (ASCC/NATO-codename 'SA-3 Goa') was, by 1961 rushed into service.[3] In order to remedy the other weakness, the long-ranged S-200 *Angara* (ASCC/NATO-codename 'SA-5 Gammon') was fielded in 1967.[4]

Concerning the S-125, the core of a missile battery was formed by four twin 5P71 launchers from which the V-600P two-stage missiles could be fired with guidance onto the target being achieved via the SNR-125 (ASCC/NATO-codename 'Low Blow') engagement radar.[5] In peacetime, the launchers, fire-control radar as well as numerous other pieces of equipment necessary to operate the system, were deployed on SAM sites of semi-permanent character. Similarly, to the earlier S-75 the S-125 was not truly mobile (vehicle mounted modernised versions of the system were post-Cold War developments) however, it was transportable on trucks and trailers. Thus, re-deployment whilst time consuming as well as laborious was possible and in fact, practised during exercises or if other circumstances called for it.

The S-200 batteries usually had six 5P72 single launchers for 5V21 missiles which had four boosters mounted in a 'wrap around' fashion on the sides of the missiles' main body with the said missiles cued onto the target by the 5N62 fire-control radar unit (ASCC/NATO-codename 'Square Pair').[6] These SAMs were deployed on fixed sites with the launchers placed semi-recessed, on concrete pads and missiles loaded from 5Yu24 transporters moving on rails. Missile storage and other supporting facilities were usually, either hardened or otherwise protected. Should the need arise, the battery could be moved from its permanent site and re-located. For this to be done, the launchers were placed on flatback trailers, missiles ferried on wheeled 5T83 transporter-loaders, the fire-control unit towed with its disassembled antennas stowed on separate trailers and the rest of the necessary equipment, similarly transported on trucks, trailers and semitrailers. Needless to say, whilst possible, the re-deployment of a S-200 SAM battery was a very cumbersome affair. As an interesting side note, the S-200 was, well into the Cold War, confused with the non-operational *Dal* (400) long-range heavy SAM (ASCC/NATO-codename 'SA-5 Griffon'). This was, in no small part, caused by the fact that missiles belonging to the latter, were sometimes paraded during celebrations of the Victory Day, October Revolution (Bolshevik coup) or on similar occasions.

Since the S-125 could engage low-flying targets but lacked range, whilst the S-200 had a long-range yet was unable to hit targets at low altitudes, once in service, both systems were paired in combined (mixed) air defence missile regiments or brigades. Thus, depending on the distance to and the flight profile of a given target, the regiment's (brigade's) commander was provided with the ability to select the SAM system which was suited best to deal with the threat at hand.

In addition, both SAM systems were continuously modernised. Known deficiencies of the S-125 were eliminated, its overall performance improved, new hardware such as the modified V-601P missile and 5P73 quadruple launchers were introduced et cetera Likewise, the S-200, underwent a similar process, in particular its range was considerably increased. If the S-200 *Angara* had a reach of 170-180km which fell short of the originally specified 200km, the S-200 *Vega* had a range of up to 240km whilst the S-200 *Dubnah* had as much as 300km (at least nominally).

Both the S-125, as well as the S-200, were based on 1950-1960s technology. This meant that that all modernisations notwithstanding, these systems had inherent limitations. In addition, combat experience in South-East Asia and in the Middle East revealed serious weaknesses of early Soviet SAM designs. For those reasons, a qualitatively new SAM system incorporating advanced technologies, had to be developed.

Accordingly, in 1969 work commenced on a new unified SAM system for Soviet Air Defence Force, the ground forces and Navy. It soon became apparent that due to diverging requirements, it would not be possible to develop such a universal system however, core technology and even some components could be shared. As a result of advances in electronics, radars and computers in particular, as well as missile technology made during the 1970s, it proved possible to introduce into PVO service, a completely new SAM system, the S-300P (ASCC/NATO-codename 'SA-10 Grumble'). From 1979, it started to replace the S-25 in defence of Moscow. At the core of the S-300P, stood the 30N6 fire-control radar (ASCC/NATO-codename 'Flap Lid') which had a passive electronically scanned array and multiple target engagement capability.[7] It worked in conjunction with the 36D6 (ASCC/NATO-codename 'Tin Shield') and 5N66M (ASCC/NATO-codename 'Clam Shell') radars.

The missiles, designated V-500, were command guided and could engage targets flying at an altitude between 25m up to 27,000m but were limited in range to 47km. They were fired from tubular canisters usually carried in fours on a single launch platform. All the systems

A V-860 or V-880 missile of the S-200 Angara/Vega/Dubna SAM system on its launcher. Still based on the same technology as S-75, this was the longest-ranged Soviet SAM of the 1960s and 1970s, capable of hurtling active radar homing missiles the size of a MiG-21, over a range of 200 and more kilometers. (Albert Grandolini Collection)

caught fire whilst the second one had its right wing blown off, with both of them crashing into the ground below. In order to gather additional data for further development of SAM nuclear warheads, a second test of this kind took place on 6 September 1961. A radar reflector suspended under an aerostat served as a target being destroyed by a nuclear missile while floating at an altitude of 22,700m.

Subsequently, no more such tests were undertaken due to the 1963 Treaty Banning Nuclear Weapon Tests in the Atmosphere, in Outer Space and Under Water. That said, some of the follow-on SAM systems were also provided with nuclear warheads: the RA-52 for the S-75 and the TA-18 for the S-200. Only the S-125 did not receive any due to that system's relatively short engagement range and altitude. Whilst a nuclear-tipped missile for the S-300 was developed, as far as it is known, it was never

components were mounted on towable trailers, semitrailers or wheeled MAZ 8x8 vehicles which facilitated relatively fast relocation from one site to another. Having arrived at a new position, it took ca. 90 minutes for the system to assume full combat readiness, that is, to be able to fire at a moment's notice. Apart from all the other new features of the S-300, this was a major advantage over earlier SAMs. In service, the S-300 was continuously modernised with new radars such as the 76N6 (also having the ASCC/NATO-codename 'Clam Shell') and the 64N6 (ASCC/NATO-codename 'Big Bird') integrated into the system. Similarly, it was augmented with new missiles which had an increased range and employed semi-active radar homing or track-via-missile guidance methods.

In case of an all-out war, the PVO's anti-aircraft rocket (missile) troops were expected to thwart strikes by strategic bombers armed with nuclear weapons or fend off large enemy aerial formations performing conventional bombardment. Such targets had to be shot down and it was imperative to provide the means which would ensure their destruction. For this reason, it was decided in 1955 to develop a nuclear warhead for the S-25 SAM system.[8] Less than two years later, the idea materialised and on 19 January 1957, a SAM with a 10kt warhead was live tested at Kapustin Yar.[9] An aiming point was provided by a radar reflector suspended under a parachute whilst two remotely controlled Ilyushin Il-28 (ASCC/NATO-codename 'Beagle') aircraft simulated an enemy bomber formation.

The missile exploded at 10,370m with the first aircraft being at about the same altitude and a distance of 570m from the point of detonation, while the second one was flying 677m lower and 787m to the rear (behind) the point of detonation. The shock wave reached the former in ca 0.7 s and the latter in ca 1.4 s. As a result, the first aircraft suffered structural damage, engine failure and

actually fielded.

Of all the Soviet SAMs described above, only the S-300 continues to serve with Russian military. The S-25 ended its service in the 1980s though the last elements of the system were dismantled in the 1990s after the USSR's demise. Meanwhile, the S-75, S-125 and S-200 were also withdrawn by the Russians however they continue to soldier on with a number of foreign operators. That said, while neither the S-25 nor the S-75 are in service with Russian air defences in their original role, large numbers of missiles manufactured and stocked for both systems are nowadays utilised by the Russians as high-speed aerial targets during air defence live-fire exercises.

Anti-Ballistic Missile Defences

As already mentioned, aside from the threat posed by manned aviation, the Soviets were also faced with the challenge posed by ballistic missiles. During the closing stages of the Second World War, the German V-2 ballistic missile proved to be unstoppable and the only practical countermeasure was to bomb its launch sites. The Soviets were studying the problem of anti-ballistic missile defences through the late1940s and early 1950s with the seriousness of the matter being recognised by top Soviet military brass. Despite many experts opining that intercepting a ballistic missile was not technically feasible, in 1956 the Soviet leadership decreed the development of an ABM system.[10]

In fact, by that time, work on such a system – designated *System A* (ASCC/NATO-codename 'Gaffer') – was already in progress with the official 'blessing' speeding things up considerably. The site chosen for the system to be deployed and tested was Sary Shagan in Kazakhstan which in those days, was a Soviet Republic and is now an independent country.[11] Incoming ballistic missile detection was

The receiver building of the Dunay-3 ABM radar in Naro-Fominsk, west of Moscow, as photographed by the KH-7 reconnaissance satellite in 1967. The system was comparable to the Pave Paws radar of the US-made Sentinel and Safeguard ABM-programmes. (US DoD)

Launcher of the A-35 ABM system. The system proved excessively expensive to research and develop. Even the leadership in Moscow was happy to limit its deployment in agreement with the USA. (Tom Cooper Collection)

enabled by a *Dunay-2* (ASCC/NATO-codename 'Hen Roost') early warning radar system which had a range of 1200km. Its 150m long receiver antenna was set-up on the shore of Lake Balkhash.

In order to facilitate an interception, a set of three RTN (ASCC/NATO-codename 'Hen Egg') and RSV-PR (ASCC/NATO-codename 'Hen Nest') radars tracked both the incoming target as well as the outgoing ABM missile so as to triangulate a point of interception. The interception itself was performed by a V-1000 two-stage missile which was fired from a launch rail mounted atop a column. The missile could reach up to 2,5000m and had a range of 300km. Its blast-fragmentation warhead contained 16,000 carbide-tungsten balls. When the warhead detonated, the said balls spread into a disc-shaped fragments field.

A series of the system's practical tests started in 1957. Progress was slowly made being marred by numerous failures. Early Soviet ballistic missiles: the R-5 (ASCC/NATO-codename 'SS-3 Shyster') and R-12 (ASCC/NATO-codename 'SS-4 Sandal') served as targets. On 2 March 1961, an almost successful test was performed – almost because due to operator error instead of the warhead (re-entry vehicle) of a R-12 ballistic missile, its main body was struck. Finally, two days later, on 4 March 1961, a R-12 missile's warhead was successfully intercepted proving that anti-ballistic missile defences were practically feasible.

Subsequently, more interceptions were performed and the effects of high-altitude nuclear explosions on the ABM system were also tested in course of Operation K (though it was not the main purpose of the tests). The operation, known also as Project K was a series of five Soviet nuclear tests in 1961-62: two explosions with a 1.2 kt yield each in 1961 and three, with 300 kt yield in 1962.[12] These were all high-altitude tests with nuclear missiles from the Kapustin Yar launch site in Russia fired across central Kazakhstan towards Sary Shagan. In the end, System A did not enter operational service remaining a prototype solely used for testing. That said, its role cannot be underestimated for it provided the foundation upon which follow-on Soviet ABM development was based.

Even before the *System A* was able to perform a successful ballistic missile interception, the Soviet leadership ordered another ABM system to be developed; one that would not just serve to prove the concept but could be operationally deployed.[13] The system designated A-35 was meant to perform exoatmospheric intercepts of incoming ballistic missiles.[14] Its main components were the *Dunay-3* (ASCC/NATO-codename 'Dog House') radar, the RKTs-35 and RKI-35 radars combo (ASCC/NATO-codename 'Try Add') and the A-350 missile (ASCC/NATO-codename 'ABM-1 Galosh'). The *Dunay-3* radar had a 200m long and 30m high transmitter as well as two 100x100m receiver arrays facing in the opposite directions, the latter co-located with the system's command and control centre.

Whilst the *Dunay-3* enabled target detection, the set of one RKTs-35 and two RKI-35 radars provided tracking and guidance, facilitating an interception. The A-350 missile was of a two-stage configuration, it had a nuclear warhead with a yield of 500kt (Western estimates were of a 2-3 megaton thermonuclear one), its length was 19.8m and it weighed 33 tons. A tubular canister with the missile was ferried on a trailer towed by a MAZ-537 heavy tractor.

With this launcher of the A-35 system in semi-open position, details of the first stage of the nuclear-tipped A-350 surface-to-air missile weighing 33,000kg can be made out. (Tom Cooper Collection)

The launcher had an unusual configuration: two cuboidal support columns placed on a trainable platform. By means of appropriate mechanisms and fittings, the canister was placed between the columns to be elevated between 60–78 degrees to set the missile on a desired trajectory when it was fired. The system's radar had a 2,500km detection range whilst the missile's range was 350km. Interestingly, the system (its launch canisters with missile mock-ups, to be more precise) was publicly shown during a parade in 1964 at a time when it was still under development. As a result, Western analysts were able to ascertain a surprising amount of information about it.[15]

Preliminary operational status of the A-35 system was first achieved in 1971.[16] Unfortunately, the system was incapable of effectively dealing with multiple, independently targetable, re-entry vehicles (warheads) whilst ballistic missiles fitted with such were by that time, already operational (e.g., the American LGM-30G Minuteman III). Since the system's practical value would be limited, it was logical to limit its deployment. Thus, launch sites for an improved A-350Zh missile were constructed at only at four locations: Klin (Klin-9), Zagorsk, Naro-Fominsk and Nudol.

A comprehensive upgrade of the system was carried out in the 1970s and 80s with among others, the modernised A-350R missiles being introduced, the *Dunay-3* radar was also modernised becoming the *Dunay-3M* in the process. Additionally, another set of large transmitter and receiver ballistic missile detection radar arrays, the *Dunay-3U* (ASCC/NATO-codename 'Cat House'), was also built. Despite all those efforts, the modernised (hence the suffix letter M) A-35M system failed to provide the desired level of ABM capability however, a new system to replace it was being developed.

In order to remedy known shortcomings of the A-35 system, work on its replacement had already started in the 1970s. The new ABM system designated A-135, was developed in Soviet times but only reached operational status after the USSR's demise, entering Russian service in 1995.[17] Its 'brain' is the *Don-2N* (ASCC/NATO-codename 'Pill Box') radar located near Sofrino in the vicinity of Moscow. Housed in a large frustum which has 130m long sides at the bottom and 90m at the top, with an overall height of 33m, the system has four circular search and track arrays as well as the same number of square arrays for missile guidance. The arrays are paired on each of the structure's sides giving the system's radars 360 degree coverage; its detection range for a target the size of an ICBM warhead is ca 3700km.

Initially, the system used two types of missiles: the 53T6 (ASCC/NATO-codename 'ABM-3 Gazelle') for endoatmospheric interception and the 51T6 (ASCC/NATO-codename 'ABM-4 Gorgon') for exoatmospheric interception but the latter has been withdrawn from service. Both types of missiles were and still are, launched from underground silos, being lowered into them inside tubular canisters.

A number of other ABM systems were in the works and have since, reached various stages of development. However, bar those described above, none became operational either in the Soviet Union nor subsequently, in Russia. Finally, these days, the Russians are developing new ABM weaponry but such are beyond the scope of matters described herein.

3
FIGHTER-INTERCEPTOR AND AIR-TO-AIR WEAPONS DEVELOPMENT

The PVO's *Istrebitelnaya Aviatsiya* (fighter aviation) had in its inventory a large number of fighter-interceptor types. Some of these aircraft were ones also used by the VVS (air force) but appropriately modified for the air defence role, whilst others were specifically designed to meet the PVO's needs. Of the latter category, two fighter-interceptors developed to defend the USSR's exceptionally long borders and immensely large area, warrant a more detailed description. Similarly, some air-to-air missiles were developed solely to arm specific PVO fighter-interceptor models whilst others, in particular short-range AAMs, were also shared with other types including tactical fighters, fighter-bombers et cetera. Finally, after a period of fascination with AAMs as the only weapons a fighter-interceptor would need, it was recognised that guns still remained useful and together with other unguided weaponry, were also provided to arm PVO's aircraft.

Soviet Mid to Late Cold War Fighter-Interceptor Development

Of the early jet types which equipped the PVO's fighter aviation,[1] some remained in service well into the 1970s, though as time went by, in lesser and lesser numbers. However, during the 1960s, the main Soviet fighter-interceptor type became the Su-9 (ASCC/NATO-codename 'Fishpot'). It was a delta-winged Mach-2 capable interceptor armed with four air-to-air missiles which served with 27 PVO fighter regiments.[2] The Su-9 was supplemented with an improved design, the Su-11 (which also had the ASCC/NATO-codename 'Fishpot') equipped with a new *Oryol* (ASCC/NATO-codename 'Skip Spin') radar and armed with two K-8 (ASCC/NATO-codename 'AA-3 Anab') AAMs. However, this type was manufactured in limited numbers entering service with only three PVO fighter regiments.[3]

Production of the Su-11 was terminated in favour of the Yak-28P fighter (ASCC/NATO-codename 'Firebar'). Meanwhile, the Sukhoi design bureau developed a new type: the Su-15 (ASCC/NATO-codename 'Flagon'). The Su-15 differed from its predecessors in that a central air inlet with a shock cone was abandoned in favour of side-mounted air intakes, which in turn, left more space for a radar and its antenna, in the nose.[4] Early production Su-15s had a pure delta wing but these were subsequently replaced by a double-delta one and the tail assembly was also modified. Initially, the Su-15 had an *Oryol* radar but later models received a *Tayfun* (ASCC/NATO-codename 'Twin Scan') set. Similarly, the armament was also modified: if early Su-15 models had only two K-8/R-98 missiles subsequently, two R-60 (ASCC/NATO-codename 'AA-8 Aphid')

Available quickly and in large numbers, MiG-19PM was armed with four RS-2U air-to-air missiles. Production ended by 1959, but dozens of PVO regiments continued flying it until the last were replaced by Su-15s, in the early 1970s. (Albert Grandolini Collection)

Su-11 was a slightly enlarged version of the Su-9, armed with two R-8 semi-active radar homing- or infra-red homing air-to-air missiles ('AA-3 Anab'), instead of the old RS-2Us beam-guiding of the first generation. (Albert Grandolini Collection)

A Su-15 interceptor of the first production version, armed with an R-8 air-to-air missile. This first variant was quickly modified through the installation of a new wing design with extended wingtips and wing fences: the extra wing area resulted in better controllability and lowered take-off and landing speeds. It retained the designation Su-15. (Albert Grandolini Collection)

The Yak-25 was the best the Soviet technology of the mid-1950s was able to offer: a relatively slow, but highly reliable, radar- and guns-equipped interceptor with a good range, with night- and all-weather combat capability. (Albert Grandolini Collection)

An early Yak-28P (notable is the short radome and lack of R-3S missiles), of the 82nd Fighter Aviation Regiment, as seen at the Nasosnaya AB by night. (via Easternorbat.com)

short-range IR guided AAMs, could also be carried. In addition, since the Soviets found out that guns were still a useful weapon complementing AAMs, provisions were made for two UPK-23-250 gun pods to be carried under the fuselage.

The final incarnation of the Su-15, with all the modifications described, was the Su-15TM. In the 1970s the Su-15 formed the backbone of PVO's fighter aviation as it equipped no less than 29 fighter regiments and two operational training units.[5] The Su-9 and Su-11 served in dwindling numbers till the year 1980 whilst the Su-15 soldiered on until the very end of the Soviet Union and even a few years beyond.[6] Finally, the Su-27 (ASCC/NATO-codename 'Flanker'),[7] one of the Soviet 'super fighters' of the 1980s, also entered PVO's service. It is a large, twin-engine aircraft with a considerable internal fuel capacity which equates to a long-range and yet still has good manoeuvrability. The Su-27 was fitted with a *Myech* (ASCC/NATO-codename 'Slot Back') radar and an IRST, has an internal GSh-30-1 gun of 30mm calibre as well as 10 hardpoints for R-27 (ASCC/NATO-codename 'AA-10 Alamo') and R-73 (ASCC/NATO-codename 'AA-11 Archer') AAMs. Having entered service in the 1980s, the Su-27 outlived the USSR and continues to serve in Russian colours as well as with a number of other operators.

Apart from numerous Sukhois, many MiGs also served with the PVO's fighter aviation. The MiG-21 (ASCC/NATO-codename 'Fishbed'), an aircraft which was so prolific as to become almost synonymous with Soviet fighters, was built in dozens of variants (to be described succinctly herein).[8] Suffice to say, it was a rather simple design with its primary armament limited to short-range AAMs. Despite the MiG-21 serving mainly with the VVS while the Su-9 being selected as the primary PVO interceptor, the former type still equipped some air defence fighter units.

The next MiG type in line was the MiG-23 (ASCC/NATO-codename 'Flogger') which was the first Soviet fighter with variable wing geometry.[9] It was un-manoeuvrable but had good acceleration. Among several fighter and fighter-bomber variants of this MiG type, the MiG-23P was a dedicated fighter-interceptor specifically equipped to work with the PVO's guidance and command systems. The *Sapphire-23* (ASCC/NATO-codename 'High Lark') radar fitted in the MiG-23 had a limited look-down/shoot-down capability. Its armament constituted of a GSh-23L 23mm cannon in a semi-recessed gun pack under the fuselage as well as two R-23 (ASCC/NATO-codename 'AA-7 Apex') and two (four if twin launchers were used) R-60 (ASCC/NATO-codename 'AA-8 Aphid') AAMs.

If the MiG-23 was adopted for the air defence role, the MiG-25 (ASCC/NATO-codename 'Foxbat') was, from the start, designed for high-speed and high-altitude interceptions.[10] The MiG-25 with its pointed nose, gaping side-mounted air intakes and high twin vertical stabilisers, made an imposing sight – especially as the aircraft was large for a fighter. Its sensors and armaments fit were composed of the *Smerch-A* (ASCC/NATO-codename 'Fox Fire') radar and four R-40 (ASCC/NATO-codename 'AA-6 Acrid') AAMs. Initially the aircraft was somewhat overestimated by foreign experts, gaining the nimbus of a 'red super fighter'. However, Lieutenant Belenko's defection in 1976 (to be described separately), exposed the MiG-25 to the West and caused a western re-evaluation of the aircraft and also, at the same time, spurred its modernisation by the Soviets.

As a result of the latter measures, MiG-25 interceptors were among others, fitted with a new *Sapphire-25* (ASCC/NATO-codename 'High Lark') radar and an IRST. Moreover, efforts to modernise the MiG-25 led to the development of the MiG-31 (ASCC/NATO-codename 'Foxhound') which will be described in more detail later. Finally, it should be mentioned that the MiG-29

A top view of a Tupolev Tu-128 (known in the West as 'Tu-28'). Thirty metres long and 40,000kg heavy when fully loaded, the type was almost the size of an airliner. It was purposely designed to search for and intercept Western bombers as far as 1,000km away from its bases in the northern USSR. At the time of its service entry in 1963, this was the most powerful interceptor of the PVO. (Albert Grandolini Collection)

(ASCC/NATO-codename 'Fulcrum'), which was one of the Soviet 'super fighters' of the 1980s, did not enter service with the PVO.

The third 'family' of fighter-interceptors serving with PVO's fighter aviation, were the Yaks. An early model was the subsonic Yak-25 (ASCC/NATO-codename 'Flashlight').[11] It is interesting to note that one such fighter used to be the personal aircraft of the PVO's fighter aviation commander General Yevgeniy Savitskiy.[12] Subsequently, the supersonic Yak-28 (bearing different ASCC/

A Yak-28P of the 174th Guards Fighter Aviation Regiment fitted with launch rails for the R-3S AAM, as seen over one of the lakes on the Kola Peninsula. (Albert Grandolini Collection)

NATO-codenames depending on its version) was developed.[13] The aircraft had two engines mounted in nacelles under each of its swept wings and was fitted with a bicycle main landing gear supported by wingtip outriggers. Of several variants including reconnaissance, strike and electronic warfare, the Yak-28P (ASCC/NATO-codename 'Firebar') was a specialised fighter-interceptor. It was equipped with an *Oryol* radar served by a dedicated radar operator who was seated behind the pilot in a tandem-type cockpit, whilst its armament compromised two K-8/R-98 and, later on, two R-3S AAMs. The Yak-28P was chosen over the Su-11 and more than 400 Yak fighter-interceptors of this type were built. Despite not being liked by its crews due to being difficult to fly, the Yak-28P soldiered on till the 1980s when it was replaced by more modern types.

A front view at a MiG-31B of the 174th Guards Fighter Aviation Regiment: formerly flying Yak-28s, the unit was one of the first to convert to the new jet, starting in January 1982. The process was completed in 1983, thus converting the 174th into the first unit ever to operate a fighter-interceptor equipped with a passive electronically scanned array radar. (Albert Grandolini Collection)

As described above, numerous fighter-interceptor types were developed or adopted for the PVO but none of them really provided a solution to the most important problem Soviet air defences were facing. Namely, how to be able to intercept enemy aircraft penetrating the USSR's exceptionally long borders and immensely large area. For this task to be handled, an aircraft was needed which, in terms of its range but also the reach of its detection and combat systems, would be like no other designed and built before. Such an aircraft which came as close to meeting those requirements as the technology available in the late 1950s and early 60s permitted, was built by Tupolev entering service under the designation Tu-128 (ASCC/NATO-codename 'Fiddler').[14] The Tu-128 was a large aircraft being in fact, the largest fighter-interceptor built to date. Indeed, the Soviets did not even consider it a fighter and consequently, units equipped with the Tupolev did not have the designation *Istrebitelniy Aviatsionniy Polk* (Fighter Aviation Regiment) but just *Aviatsionniy Polk* (Aviation Regiment).

The Tu-128 had a range of ca 2,500km, was fitted with a *Smerch* (ASCC/NATO-codename 'Big Nose') radar and armed with four R-4 (ASCC/NATO-codename 'AA-5 Ash') AAMs becoming the first Soviet interceptor which had a BVR engagement capability. Crewed by two men (pilot and radar operator), it was equipped to work with ground-based command and control systems as well as the Tu-126. Originally, plans called for as many as twenty-five Tu-128 regiments but in the end, it served with just six.

Operating it presented numerous challenges including training for it was a completely different aircraft to fly than a 'normal' fighter. Consequently, pilots were trained on *Il-14* (ASCC/NATO-codename 'Crate') transports and *Tu-124* (ASCC/NATO-codename 'Cookpot') passenger aircraft till finally a dedicated Tu-128UT trainer became available. The Tu-128 stayed in service into the 1980s, with remaining airframes being physically destroyed in the next decade, bar a few ones left for museum displays.

Meanwhile, advances in technology permitted building a new aircraft which constituted a huge leap in the PVO fighter aviation's long-range interception capability. Its development started as a modernisation project of the MiG-25 (see previously), but it evolved into a completely new design although it showed some external resemblance to the former. The said aircraft is the MiG-31 fighter-interceptor.[15] It is fitted with a *Zaslon* (ASCC/NATO-codename 'Flash Dance') phased array radar, which together with four long-range (120km for the early version) R-33 (ASCC/NATO-codename

The MiG-31 represented the ultimate Soviet interceptor: the requirement for a long-range, powerful weapons system with a heavy punch, resulted in a big aircraft with poor manoeuvrability, but also one that was fast, equipped with a very capable radar and fire-control system and carrying four R-33 long-range air-to-air missiles as its primary armament. The latter could be enhanced by two R-40s, or – as visible on this example intercepted over the Baltic Sea in the late 1980s – short-range R-60s, useful for self-defence purposes. (Albert Grandolini Collection)

AIRCRAFT AND MISSILE DESIGNATION SYSTEMS

Generally, Soviet designations for military aircraft, missiles and weapons systems were all considered state secrets: therefore, not only were very few ever made public before the 1990s, but there were actually, two designation systems, of which one was 'public', and the other 'secret'. The 'secret' system was relatively complex and consisted of three parts: because it remains largely unknown, even in the former USSR, a discussion of it will be omitted. The 'public' system – which can be considered as having resulted in the 'real' designations – was, at least for insiders, 'quite simple': when it comes to missiles, it started with the prefix R-. Further to that, every missile was a part of the weapons system, which was prefixed with K-, whilst each missile also had its factory designation, prefixed by *Izdeliye* ('Model'). The prefix was followed by a one or two-digit sequential number denoting the specific type of the missile (where odd numbers were preferred, because of a strong Russian belief that odd numbers are lucky). Depending on its further development, the designation was then followed by a suffix denoting the modification standard.

Because the mass of them remained unknown in the West until the 1990s, designations of Soviet aircraft and air-to-air missiles have caused quite some confusion and not a few controversies, during the Cold War. In attempt to sort out the resulting chaos, the West thus referred to Soviet fighter aircraft and their armament by a sequential code assigned by the Air Standardisation Coordinating Committee (ASCC). This body was established during the Second World War, and – between others – was responsible for assigning reporting names to unknown types of German, Italian, and Japanese aircraft.

Initially, it comprised representatives of the Australian, British, Canadian, New Zealand and US armed force. When the North Atlantic Treaty Organization (NATO) came into being, the ASCC's system of code-names and reporting names was retained, but these became best-known as 'NATO reporting names' in the public.

As a result of these multiple systems, several weapons became known in public under entirely different designations. Perhaps the best example was the R-3 missile: originally little else but a reverse-engineered copy of the US-made GAR-8 (or, since 1962: AIM-9B Sidewinder), its first sub-variant had the factory designation *Izdeliye* 310 and was the part of the K-13 weapons system. Because the latter became known first outside the USSR, the weapon was frequently referred to – and is still often referred to – as the 'K-13'. Actually, the version manufactured in biggest numbers, most widely applied in the USSR, and most widely exported, was the slightly improved R-3S (Izdeliye 310A), which was a part of the K-13A or K-13S weapons system. The following table provides the most important details for Soviet-made fighter-interceptors and air-to-air missiles relevant for the PVO from 1960 until 1991.

Table 5: Airborne Early Warning Aircraft & Fighter-Interceptors of the V-PVO, 1960–1991

Soviet Designation	ASCC Designation
A-50	Mainstay
MiG-17, MiG-17F	Fresco
MiG-19S, MiG-19P, MiG-19PM	Farmer
MiG-21F-13/S/SM	Fishbed
MiG-23M/MLA/P	Flogger
MiG-25P/PD/PDS	Foxbat
MiG-31	Foxhound
Su-9	Fishpot
Su-11	Fishpot
Su-15, Su-15TM	Flagon
Su-27P	Flanker
Tu-126	Moss
Tu-128	Fiddler
Yak-25, Yak-25M	Flashlight
Yak-28P	Firebar

Table 6: Air-to-Air Missiles of the V-PVO, 1960–1991

ASCC Designation	Soviet Designation	Aircraft Type	Weapons System & Notes
AA-1 Alkali	RS-1, RS-2, R-55, RS-1U, RS-2U, R-55, R-55M	MiG-19PM, Su-9	beam-guided
AA-2A Atoll	R-3S	MiG-21, Yak-28P	K-13; IR-homing
AA-2C Atoll	R-3R	MiG-21, Yak-28P	K-13; SAR-homing
AA-2D Atoll	R-13M	MiG-21, MiG-23M	K13; IR-homing
AA-2-2 Advanced Atoll	R-13M1	MiG-21, MiG-23M	K-13; IR-homing
AA-3 Anab	R-8, R-98R, R-98T, R-98MR, R-98MT	Su-11, Su-15, Yak-28P	K-8, K-8M, K-98
AA-3 Advanced Anab	R-98MR, R-98MT	Su-15TM	SAR- and IR-homing, respectively
AA-4 Awl	R-9, R-38	-	K-9; never entered operational service
AA-5 Ash	R-4R, R-4T, R-4MR, R-4MT	Tu-128	K-4; IR-homing (T) and SAR-homing (R)
AA-6A Acrid-A	R-40R	MiG-25P	K-40; SAR-homing
AA-6B Acrid-B	R-40T	MiG-25P	K-40; IR-homing

Table 6: Air-to-Air Missiles of the V-PVO, 1960–1991 (continued)			
AA-6C Acrid-C	R-40RD	MiG-25PD/PDS, MiG-31	K-40; SAR-homing
AA-6D acrid-D	R-40TD	MiG-25PD/PDS, MiG-31	K-40; IR-homing
AA-7A Apex-A	R-23R	MiG-23M	K-23; SAR-homing
AA-7B Apex-B	R-23T	MiG-23M	K-23; IR-homing
AA-7C Apex-C	R-24R	MiG-23M/MLA/P	K-23; SAR-homing
AA-7D Apex-D	R-24T	MiG-23M/MLA/P	K-23; IR-homing
AA-8 Aphid	R-60, R-60MK	MiG-21, MiG-23, MiG-31, Su-15, Su-15TM	IR-homing
AA-9 Amos	R-33, R-33S	MiG-31	K-33
AA-10A Alamo-A	R-27R	Su-27P	K-27; SAR-homing
AA-10B Alamo-B	R-27T	Su-27P	K-27; IR-homing
AA-10C Alamo-C	R-27ER	Su-27P	K-27; SAR-homing
AA-10D Alamo-D	R-27ET	Su-27P	K-27; IR-homing
AA-1 Archer	R-73, R-73M1	Su-27P	IR-homing

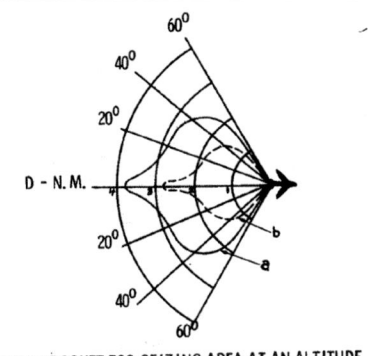

DIMENSIONS OF R-3S ROCKET TGS SEIZING AREA AT AN ALTITUDE OF 50,000 FT
(a) LONG RANGE BOMBER
(b) FRONT LINE FIGHTER
FIGURE 15

TABLE I
THEORETICAL ROCKET-LAUNCHING RANGES

OVERTAKE RATE ~ Kts	55	110	100	215	270
	Minimum Theoretical Permissible Launching Ranges				
	1800	2000	2200	2400	2600
ALTITUDE (ft)	RANGES ~ FT.				
3,300	5,900	6,900	8,200	9,400	10,500
16,500	7,800	9,200	10,800	12,500	14,000
33,000	14,000	16,400	19,000	22,000	24,000
49,000	24,000	26,000	28,000	30,000	32,000
66,000	26,000	27,500	30,000	31,000	33,500

NOTE: The underlined distances are those which exceed the distance from which aiming begins, (radar or visually). Interceptor speed for altitude of 66,000 kts is 1080 kts, for the remaining altitudes, 485-595 kts.

What was it all about: a diagram from a translated version of the Soviet tactical manual for MiG-21F-13 and MiG-21PF interceptors, depicting the engagement envelope of the R-3S air-to-air missile ('AA-2 Atoll') against a target underway at an altitude of 50,000ft (15,240m). For most of the 1960s, the PVO expected to operate its interceptors against bombers underway at similar or higher altitudes, in form of a stern-conversion, followed by a missile attack from ranges of (in the case of the R-3S) 1,800 to 2,600m. Effective engagement envelopes for missiles like R-8/98 and R-4 were about double that. (Tom Cooper Collection)

TABLE IV
Probability of an Interceptor Being Hit During One Complete Attack

Pullout Distance ~ (ft)	Overtake Velocity ~ (kts) →	Fire from B-52 or B-58 type Aircraft				Fire from B-47 type Aircraft			
		55	80	110	160	55	80	110	160
2600		0.42	0.34	0.27	0.10	0.25	0.19	0.15	0.11
2000		0.54	0.44	0.36	0.26	0.34	0.26	0.21	0.16
1500		0.68	0.58	0.48	0.35	0.45	0.36	0.27	0.21

The data in the table shows that with the increase in rate of approach from 55 to 160 knots, the probability of being hit decreases by almost one-half.

Generally, PVO pilots were not expected to engage in air combats with enemy fighters. On the contrary, theoreticians of the General Staff of the Soviet Armed Forces expected the tail-mounted barbettes of US-made bombers to represent the biggest threat for PVO's interceptors. This table from the same tactical manual, is discussing the probability of an interceptor being hit by self-defence armament of primary bombers of the US Air Force (USAF) of the early 1960s, during a stern-conversion attack. (Tom Cooper Collection)

'AA-9 Amos') AAMs, carried semi-recessed under the fuselage, forms the core of the fighter's combat capability.

In addition to the radar, an IRST was also fitted. Even more importantly, the MiG-31 is equipped to receive tactical information from ground-based and airborne (A-50 aircraft) air surveillance platforms, as well as to exchange such with other MiGs: all this greatly boosting its effectiveness in the air defence interceptor role. The main armament can be augmented by additional AAMs carried on underwing pylons: either two R-40 or four (using twin launchers) R-60 and the aircraft is also armed with a GSh-6-23 six-barrel 23mm cannon. As a result of the complexity of the aircraft's system, a two-man crew proved necessary in order to fly and operate it. The MiG-31 entered PVO's fighter aviation service in the 1980s being arguably, what Soviet AD needed throughout its existence and eventually received – but only in the USSR's final decade. However, the end of the Soviet Union did not end the MiG-31's career, for it continues to serve with Russia and Kazakhstan, remaining one of their most valuable AD assets.

Soviet Air-to-Air Missile Development

Before describing individual Soviet air-to-air missile types, it needs to be pointed out that the Soviets viewed AAMs as a part of a *kompleks*, that is a complex or system which integrated the missile with an interceptor's on-board radar and both, with the aircraft which carried them into a functionable entity meant to engage aerial targets.[16]

By 1960, the Soviets had two AAMs in active service: the RS-1 and RS-2 (ASCC/NATO-codename 'AA-1 Alkali') and the R-3 (ASCC/NATO-codename 'AA-2 Atoll').[17] The former was radar (beam-riding) guided while the latter was infra-red homing and indeed: a copy of the American GAR-8/AIM-9 Sidewinder missile. Since fighter types in Soviet service such as the Su-9 and the MiG-21 were, as a rule, fitted specifically to carry one of the missile types, the limitations of their guidance methods meant that these aircraft were also limited in their ability to engage aerial targets.[18] In order to address this, an IR guided version of the RS.2 was developed and designated R-55 (this retained the ASCC/NATO-codename 'Alkali').

Furthermore, a semi-active radar homing version of the R-3S was developed in form of the R-3R ('AA-2C Atoll'). Thus, it became possible to arm the Su-9 as well as later models of the MiG-21 with a combination of radar and IR guided missiles. This became the Soviet's standard practice: that is, to develop both a radar guided as well as an IR guided version of a given AAM type and to arm their fighter-interceptors with missiles paired so as to have an AAM combat load combining radar and IR guided ones.

During the late 1950s, a number of new Soviet medium-range AAMs were being worked on to arm fighter-interceptors which were about to enter PVO fighter aviation's service. Of these, the R-8 and its modifications would arm no less than three interceptor types. First, combined with the *Oryol* radar, it constituted the core of the K-8 weapons systems of the Su-11 and Yak-28. It had a range of only 12km and was handicapped by many limitations of its engagement envelope but could boast a 40kg blast-fragmentation warhead. An improved version of the missile, initially designated R-8M, subsequently re-designated to the R-98, was developed for the Su-15 fighter-interceptor.

It is fitting at this point to explain that basic designations were, in the course of a missile's development, supplemented by suffix letters: usually M for modernised, R for radar guided and T for IR guided – this also applying to most other AAMs described herein. The final version of the missile was the R-98M (MR and MT) compatible with the Su-15TM's *Tayfun* on-board radar. Its maximum range was 24km, which was double that of the original K-8 and it had a much wider engagement envelope. This weapon was given the ASCC/NATO-codename 'Advanced Anab'. As described, the K-8/R-98 missile family was prolifically arming several fighter-interceptor types from the 1960s to the early 90s but the withdrawal of the Su-15, not long after the USSR's demise, also meant the withdrawal of its last member the R-98M missile.

Meanwhile, the R-4 AAM was being developed, specifically designed to work with the Tu-128 interceptor's *Smerch* radar and meant to be its main and only armament. It was a large weapon by any standard, being just over 5.5m long and weighting ca 500kg of which 43.4kg accounted for the warhead. The missile's size was necessary to accommodate a powerful rocket motor with a sufficient amount of fuel but also to house its guidance electronics which in those times, were large and heavy. All this was needed to provide the R-4 with a range of 25km. It was the first operational Soviet AAM with such a relatively long reach that was vital to the Tu-128's long-range interception capability. Due to technological progress, the R-4 fell into obsolescence (upgrades notwithstanding) but remained in service with the Tu-128 till the type was withdrawn.

Another large and heavy Soviet-era AAM was the R-40. It had a weight of 470kg including a 38kg warhead and a range of 30km (specific data for various sub-types of the missile may differ). Excluding prototypes, research projects and the like, the R-40 was with a length of 6.29 metres the largest ever AAM to have entered

An R-8R air-to-air missile ('AA-3 Anab'), as seen installed on Su-15 of the 166th Guards Fighter Aviation Regiment while standing the quick reaction alert (QRA) at the Sandar AB, outside Marneuli, in the early 1970s. (via Easterorbat.com)

A 5.2m (17ft 1in) length and with a wingspan of 1,300mm (4ft 3in), and weighing 493kg on launch, at the time of its service entry in 1963, the Bisnovat R-4 ('AA-5 Ash') was the biggest air-to-air missile in world-wide service. Such size was badly necessary to accelerate it to at least Mach 1.6 while carrying a 53kg-heavy warhead over the maximum range of 15–20km: performances necessary to knock-down a bomber the size of the B-52 with a single blow. (Albert Grandolini Collection)

only became available when the MiG-31 with its *Zaslon* phase array radar and the R-33 AAM combo, reached operational status in the 1980s. This radar/missile system is highly capable as it can simultaneously track 10 targets and guide missiles on four of them. Moreover, it has the capability to engage small and low-flying ones such as cruise missiles. Since the MiG-31 not only remains in Russian and Kazakh service but is in fact, one of their most valuable AD assets expected to soldier on for years to come, the R-33 missile is by extension, also sure to serve well into the future.

Apart from medium and long-range AAMs, the Soviets also developed new short-range IR guided ones. Leaving aside their use by tactical fighters and fighter-bombers, these were meant to supplement

operational service anywhere. The missile was compatible with the *Smerch-A* and subsequently the *Sapphire-25* radars of the MiG-25 fighter-interceptor being the type's main armament. After the MiG-25's withdrawal, an upgraded version of the missile continued its service as the MiG-31's secondary armament but has since been retired.

Efforts to provide a potent medium-range AAM for the MiG-23 fighter compatible with its *Sapphire-23* radar, resulted in the R-23 missile. The said AAM had a range of 27km and a 26kg warhead yet despite a protracted development time, the R-23 missiles were far from perfect. In order to improve the MiG-23 missile armament's capabilities, in particular, target acquisition and hit probability, as well as general reliability, a much-improved AAM – the R-24 – was designed. The newer missile shared with its predecessor, the general configuration and appearance as well as the ASCC/NATO-codename 'AA-7 Apex'. Currently, the R-23/24 remain in service with a few foreign operators still flying the MiG-23 fighter, but no longer with Russia.

For the purpose of arming the next generation of Soviet fighters, the MiG-29 and Su-27 in particular, a new medium-range AAM, the R-27, was developed. Among many advanced features it incorporates, is a quick launch capability. It is a significant improvement over earlier Soviet medium-range AAMs which usually had a relatively long firing sequence which is a serious drawback in a combat situation when time is of the essence. Of the two fighter types mentioned, only the Su-27 served with the PVO's fighter aviation. Accordingly, the R-27 missile went into the PVO's inventory as a part the Su-27's weapon system mated with the fighter's *Myech* on-board radar. The R-27, despite newer AAM's being developed and fielded, still serves with the Russian as well as a number of foreign air arms.

Considering the PVO fighter aviation's primary mission of defending the huge Soviet landmass and long borders, some of the AAM types described were less useful whilst others were more. However, none of them had what was needed to be up to the task. This

the main missile armament of fighter-interceptors. They were also to provide them with a weapon better suited for close range combat, should any take place, than the primary AAM types they were armed with. The first new weapon of this kind was the R-60, which in terms of manoeuvrability but especially seeker quality (and thus, also hit probability), was a major advance over earlier Soviet missiles, in particular over the most proliferous Soviet IR guided short-range AAM the R-3/K-13. However, the R-60 was not without shortcomings itself: especially its short-range and relatively small warhead of just 3–3.5kg were serious limitations of this weapon. These were addressed with the development of the R-73, which had a better seeker and greater manoeuvrability whilst at the same time, also boasted a considerably longer effective range and a more potent 7.4kg warhead. Both missile types remain in widespread service throughout the former Soviet Union and the rest of the world.

Finally, a number of AAMs were being actively developed by the Soviets but either failed to enter service or did so only after the end of the Soviet Union with the R-77 missile (ASCC/NATO-codename 'AA-12 Adder') being an example of the latter. However, describing such is beyond the limits of the present analysis.

Guns, Rockets, and Bombs

Whilst air-to-air missiles became the primary armament of fighter aircraft, theories that AAMs would make guns completely redundant were proven wrong by experience. Accordingly, the Soviets designed a number of cannons to arm their fighters. The PVO's fighter fleet was no exception and many (though not all) Soviet fighter-interceptors, including heavy ones, were also provided with this type of armament.

The most widely used Soviet aviation gun became the GSh-23.[19] It is a twin-barrelled weapon of 23mm calibre in which the firing action of one barrel operates the mechanism of the other, this being the so-called Gast principle.[20] The main advantage of such an arrangement is that it provides a much faster rate of fire for lower mechanical wear, compared to a single-barrel weapon. Since many

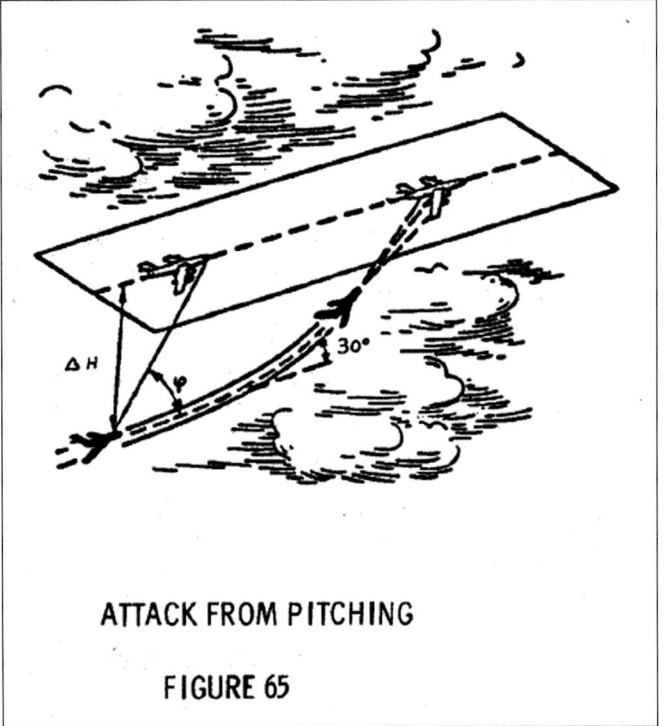

ATTACK FROM PITCHING

FIGURE 65

One of the instructions for an attack with unguided rockets against a higher-flying target, from the tactical manual for the MiG-21F-13/PF issued by the General Staff of the Soviet Armed Forces in the early 1960s.

series of fighters, with one being attached under the fuselage. Follow-on MiG-21 models and the MiG-23 fighters did not carry gun pods but instead had a ventral 'gun pack' installed semi-recessed in the fuselage underside.

As the need for an internal gun was finally recognised, late Soviet-era fighter aircraft were designed with one. Of the types serving with PVO units, the MiG-31, was armed with the 23mm GSh-6-23 cannon whilst the Su-27 had a 30mm GSh-30-1 piece. The former is a six-barrelled rotary (i.e., Gatling-type) gas operated weapon, which, by means of a rotating barrel assembly, delivers sustained fire at a very high rate of 6000 rounds per minute. However, in service the gun was plagued by reliability and safety issues which resulted in its use being severely restricted. In contrast, the GSh-30-1 was a traditional design with a single-barrel utilising short recoil action to operate. This makes for a relatively light and compact weapon with a slower rate of fire which has a number of advantages: reduced barrel wear as well as ammunition expenditure and a high degree of reliability.

Guns were not the only unguided weapons PVO's fighter-interceptors were armed with. The 57mm S-5 rockets held in UB-16 and UB-32 pods (containing 16 and 32 projectiles respectively) were also sometimes carried and all pilots trained for deploying them in air combat.[21] There was a certain logic behind that: AAMs could be rendered useless by means of countermeasures (decoys and jamming) whilst guns might prove insufficient in destructive power against a target such as a Boeing B-52 Stratofortress bomber. Yet, since a strategic bomber was likely to carry nuclear weapons, it had to be destroyed and a salvo of unguided rockets could accomplish just that.

Soviet fighter types were designed without an internal gun, several gun pods for the GSh-23 were developed. In particular, the UPK-23-250 pod which contained the gun and 250 rounds of ammunition was used to arm Sukhoi fighter-interceptors. It was tested, but not operationally carried, on the Su-9. However, it subsequently armed the Su-15 with two such gun pods suspended ventrally under the fuselage, which gave the aircraft considerable firepower if the situation warranted the use of a gun. Additionally, the G-9 pod (sometimes referred to as a gondola) for this gun having a capacity of 200 rounds, was developed and used to arm the MiG-21PF/PFM

Fortunately, American bombers never showed up over the USSR but foreign aerostats did with the rockets sometimes being used against them (more on this in Chapter 6). In addition, the poorly thought out PVO reform of late 1970s early 80s (see Chapter 1), saw the transfer of a number of PVO fighter-interceptor units to the VVS. Naturally, this gave them a more tactical role, possibly requiring in war time to perform ground attack and strike missions. Unguided

The principal target of the V-PVO through the 1960s, 1970s, and 1980s were Boeing B-52 Stratofortress bombers of the Strategic Air Command (SAC), USAF. Armed with several thermonuclear weapons, these were expected to strike all major urban centres in the USSR. This example is seen while standing alert, with its crew running to man the jet for a simulated scramble. (Albert Grandolini Collection)

Another potential target, the interception of which was trained intensively by PVO pilots of the 1960s and 1970s, was the Convair B-58 Hustler bomber. The SAC operated three wings of these, all armed with thermonuclear bombs and the Soviets expected them to penetrate their airspace at speeds of Mach 2 – and trained their pilots to intercept – at altitudes up to 29,870m (98,000ft)

rockets were an appropriate weapon for the former, whilst bombs for the latter. However, aircraft designed specifically as fighter-interceptors, were ill-suited for those type of missions, though bomb carriage was in fact, tested on the Su-9 and the Su-15. Luckily for all those concerned, this never progressed beyond trials and as already described, above the PVO reform was in the end, also reversed.

4
STRATOSPHERIC SHENANIGANS

Few things epitomised Cold War rivalry, especially its aerial dimension, like US high-altitude reconnaissance missions and the corresponding Soviet efforts to intercept them. After the well-known 1 May 1960 downing of the U-2 deep inside the USSR, American and other western high-altitude reconnaissance aircraft stopped venturing over the Soviet heartland. That said, peripheral missions, sometimes ones skirting Soviet airspace or at the very least coming close to it, continued. With the fielding of the SR-71 strategic reconnaissance aircraft, the United States gained an asset allowing it to perform such practically with impunity, at least up to a point. Naturally, the PVO challenged these missions and towards the end of the Soviet Union, had demonstrated a certain capability to effectively oppose them if ordered to do so.

Deja vu
As stated, U-2 missions over the USSR ceased, though technically speaking, they were merely halted on the order of President Eisenhower, an order his successor John F. Kennedy upheld but one which, theoretically, could have been countermanded at any given time.[1] Thus, the U-2s were never again sent over the Soviet Union deliberately, yet they still ended up in Soviet airspace causing an incident. In order to explain how this came about, it is necessary to look to the other side of the globe. Namely, in the fall of 1962, all eyes were turned on Cuba where a crisis erupted between the USA and the USSR, caused by the deployment of Soviet nuclear missiles to that island. Those historical events are well-known and for this reason, relating them here at length can be dispensed with. Suffice to say, on 27 October 1962, a U-2 flown by Major Rudolf Anderson, was shot down over Cuba by a Soviet-crewed S-75 (SA-2) SAM[2] and tension reached such a point that a nuclear war became a realistic possibility. The actors involved were conscious of the gravity of the situation and made efforts to de-escalate.

In the end, they succeeded averting war, yet the climb-down was, for various reasons, not easy to bring about, especially as another aerial incident involving the violation of Soviet airspace, took place. Namely, on the very same day as Major Anderson was shot down, another U-2 with Captain Charles W. Maultsby, was flying an air sampling mission from Alaska. Over the North Pole, the American pilot made a navigational error due to the unreliability of the compass in those parts of the world, as well as the inability to take a proper celestial navigation fix because of the aurora borealis (northern lights). As a result, the U-2 strayed way off course and flew into Soviet airspace. Alarmed by the intrusion, the Soviets scrambled four fighter aircraft – two MiG-19s and the same number of MiG-17s – which trailed the US aircraft but failed to intercept as it flew at an altitude too high for them to succeed in their endeavour. In response to Soviet fighter activity, the Alaska Air Defence Command scrambled a pair of F-102 interceptors armed with 'Falcon' air-to-air missiles which had nuclear warheads (*sic!*). As interesting as air combat involving the use of nuclear AAMs would have been, it is most fortunate such was avoided because there was

Although rapidly growing obsolete, large numbers of MiG-17 and MiG-17Fs were still forming the backbone of the PVO throughout the 1960s. This example was photographed while intercepting a Boeing RC-135 of the USAF in 1971. (US DoD)

no need for the American fighters to intervene as the Soviets were unable to intercept the high-flying U-2.

Thus, Captain Maultsby's U-2 exited Soviet air space, unmolested but out of fuel, it was able to glide for nearly an hour before landing at Kotzebue on the western tip of Alaska. As an interesting side note, his flight of 10 hours and 25 minutes was the longest ever recorded for a U-2. Fortunately, the whole incident had no repercussions other than wrecking the already strained nerves of all those involved. This is illustrated by the un-presidential, yet all things considered, understandable remark by President Kennedy that there is always some son of a bitch who doesn't get the word.[3] The events described could bring about a feeling of *deja vu* since they strongly resembled previous incidents involving the U-2. However, it was the last one of this kind for a new player was about to enter the game of high-altitude reconnaissance, ratcheting it up to an entirely new level, both literally and figuratively.

A New Player Enters the Game

Even before the U-2 became operational, ideas about its replacement were already being contemplated.[4] A lengthy discussion of this matter took place, followed by the evaluation of various designs. On 11 February 1960, a contract was signed with Lockheed, its subject being a high-speed, high-altitude reconnaissance aircraft for the CIA. The said aircraft became the A-12: a Mach 3.29, 90,000 feet (27,400m) capable single-seat (crewed only by one pilot) reconnaissance aircraft, with the entire project being codenamed 'Oxcart'. As far as known, the A-12 never flew a mission reconnoitring the USSR, although the idea to fly one along the Soviet-Finnish border and down the Baltic coast, was mulled over. Nothing came of that so instead, from 1967 onwards, the A-12 sortied over North Vietnam and North Korea. In all, just 29 operational missions by the A-12 were flown, after which the aircraft was phased out in 1969. The abandonment of the A-12 did not result from any shortcomings of the aircraft itself. Rather it was caused by a combination of fiscal pressure and competition between aerial reconnaissance programmes of the CIA and the USAF, with the latter coming out on top.

The A-12 spawned a series of advanced aircraft projects based on a common airframe. Among others, USAF officials showed interest in the RS-12 reconnaissance/strike variant which subsequently evolved materialising as the SR-71 strategic reconnaissance aircraft.[5] The aircraft became colloquially known as the Blackbird, reportedly earning this nickname because of external coating with a high-emissivity black paint for improved heat radiation which reduced the thermal stresses on the airframe.[6] In December 1962, the USAF ordered six (more later) such aircraft, which were originally designed to conduct high-speed, high-altitude reconnaissance of enemy territory after a nuclear strike.[7] It may be of interest to note, that as revealed by a former SR-71 pilot Colonel Rich Graham, the titanium used for the SR-71's airframe, was primarily sourced through intermediaries from the Soviet Union.[8] The main differences between the A-12 and the SR-71, externally quite similar, was that the latter was designed with emphasis on multi-sensor data collection and that it had a two-man crew. Apart from the pilot, who flew the aircraft, a second crewman was seated behind him in the SR-71's tandem-type cockpit. He was the reconnaissance systems officer tasked with operating the aircraft's surveillance systems and equipment, as well as directing navigation according to a given mission's flight path. Additional sensors and a second crewmember increased the weight and thus, correspondingly reduced the new aircraft's performance vis-à-vis its predecessor. Thus, the SR-71's maximum recommended cruise speed for normal operations, stood at

An SR-71 refuelling from a Boeing KC-135 Stratotanker during a mission in 1983. (USAF)

Mach 3.17 whilst a typically operational altitude was between 70,000 and 85,000 feet (ca 21,300 to 25,900m).[9]

The CIA's A-12 was withdrawn in 1969 but before that happened, the USAF's SR-71 entered service conducting its first operational sortie on 21 March 1968.[10] The SR-71s were sent on missions over North Vietnam and elsewhere in the Far East, as well as in the Middle East and North Africa. Since the Americans had such a capabilities-wise unique aircraft, it was just a matter of time before the SR-71 would be used to reconnoitre the United States' primary Cold War rival – the Soviet Union.

On the 'Red Empire's' Fringes

There were numerous locations in the USSR which were of interest to US intelligence. However, since for reasons already explained, direct overflights, especially of deep penetrating nature, were out of the question so the Americans had to be content with reconnoitring from outside of Soviet airspace sites, located in border areas utilising stand-off viewing. In the Far East, these were primarily Petropavlovsk, on the eastern coastline of the Kamchatka Peninsula and Vladivostok, further to the south.[11] Both were important naval bases and there were also many other military installations in their vicinity. Similarly, in the Soviet Far North: Murmansk, on the Barents Sea, but also Zapadnya Litsa, Vidyayevo, Gadzhievo, Severomorsk and Gremikha, attracted US attention due to the naval bases located there.[12] Of particular interest were Soviet nuclear submarines armed with sub-launched ballistic missiles, other nuclear and conventional submarines, as well as remaining naval assets and everything else of military or intelligence value.

In addition, 'Blackbirds' also flew missions along the Soviet Baltic coast (nowadays for the most part belonging to independent Baltic countries) and swung around the Swedish island of Gotland setting for a return flight in the opposite direction.[13] Blackbird missions along Soviet borders started in the 1970s and continued through the 1980s. On numerous occasions, the SR-71s flew close to the Soviet airspace but as far as it is known, did not actually penetrate it. That said, it was claimed (though not confirmed), that in the course of a mission flown in the vicinity of Vladivostok on 27 September 1971 to collect radar signals of the S-200/SA-5 SAM, a Blackbird flown by Colonel Robert (Bob) Spencer and Colonel Richard (Butch) Sheffield did enter Soviet airspace.[14] However, assertions that a SR-71 overflew Moscow can be definitely dismissed.[15]

To the extend it can be ascertained from available sources, most SR-71 missions were flown by single aircraft, however at times, two 'Blackbirds' were involved. A particularly noteworthy mission conducted by two SR-71s, took place off the coast of Petropavlovsk in early 1980: heading directly towards each another and maintaining a reciprocal course, the 'Blackbirds' passed one another head on with ca. 5,000ft (a little more than 1,500m) vertical separation, at a closure rate of over Mach 6.[16] This was not performed just for the sake of it but in order to provoke Soviet radar activity which was recorded by the SR-71s' on-board systems and by a RC-135 also sent to the area, for subsequent analysis. As a matter-of-fact, the SR-71s missions were at times, conducted in coordination with other intelligence-gathering assets, in particular aircraft such as the aforementioned RC-135, but also British 'Nimrods' or Norwegian 'Orions'. The purpose was for those other aircraft to gather additional intelligence material in particular electronic emissions. A SR-71 mission could result not only in increased Soviet radar activity, but also additional communications as well as fighter aircraft scrambles. These also yielded plenty of electronic intelligence material to be recorded and analysed.

Hunting the Blackbird

The SR-71 was never shot down and, as far as is known, was not even fired on by the PVO; however, that does not mean that preparations to that end were not made. Of the assets available to Soviet air defences', high-altitude SAMs and fighter-interceptors were the means which theoretically, could facilitate the downing of a Blackbird.

Concerning surface-to-air missiles, the SR-71 was on a number of occasions, fired upon by North Vietnamese, North Korean and Libyan SAMs but never effectively. However, an A-12 was in one instance damaged by a SAM. Namely, on 10 October 1967, while conducting a mission over North Vietnam an A-12 with pilot Dennis Sullivan at the controls, it was fired on by two S-75/SA-2 (the Vietnamese had no other) SAM sites which together launched no less than six missiles. At the time of coming under attack, the aircraft was flying with a speed of Mach 3.1 at an altitude of 84,000 feet (ca 25,600m).

A post-flight inspection revealed a piece of metal lodged against a support structure of the right wing fuel tank which had penetrated the right wing's underside, passing through three layers of titanium. It appears the piece of metal was not a fragment of the warhead but rather, debris from the missile's body.[17] That was of secondary importance as what mattered was that the A-12 was hit. Consequently, even an aircraft flying at an altitude of about 80,000 feet (approximately 24,000m) and with a speed of Mach 3, could not consider itself to be completely safe. Thus, a Soviet S-200/SA-5 SAM had a chance of shooting-down a SR-71 but arguably, for this to happen, everything would have had to perfectly align for the Soviets whilst the Americans would have had more than their fair share of bad luck. In any case, this remains moot, for a Blackbird was never fired on in anger by the PVO's SAMs.

Of the fighter-interceptor types in service with the PVO, the MiG-25 was the first one which could have had a theoretical chance of engaging a SR-71. As a matter-of-fact, Blackbird activity in the Far East prompted the PVO's fighter aviation's commander General Yevgeniy Savitskiy, to verbally instruct pilots of the MiG-25-equipped 530th IAP (the same unit from which Lieutenant Victor Belenko would defect with his aircraft), to shoot a SR-71 down if a suitable opportunity presented itself, with the wreckage likely to fall into Soviet territorial waters.[18] However, while being de-briefed by the US intelligence about his defection, Belenko stated that due to the limitations of the aircraft itself as well as its target acquisition and weapon systems, a MiG-25 was in practical terms, incapable of shooting-down a Blackbird.[19]

It appears SR-71 pilots did not feel particularly threatened by the MiG-25. For example, Colonel Rich Graham witnessed a Soviet interception attempt by a trio of MiG-25s while he was piloting a SR-71 at an altitude of 75,000 feet (23,000m) flying with a Mach 3 speed in the vicinity of Petropavlovsk some 100 nautical miles (ca 160km) off the USSR's coast.[20] All three Soviet fighters flew by right underneath his mount and Colonel Graham did not assess them as constituting a serious problem for him. Attempts to intercept the 'Blackbirds' were also made by MiG-25s of the 787th IAP based at Eberswalde-Finow in the former German Democratic Republic (East Germany).[21] Starting in 1980, a new alarm call *Yastreb* (hawk) was introduced which meant that a SR-71 was approaching. The issuing of this particular alarm call usually prompted the scrambling of 787th IAP's MiG-25s. When intercepting a Blackbird over the Baltic, the MiG-25s would be flying a wide curve on a parallel course so as to assume a position at an altitude of 22,000m and 3,000m distance to the rear of the intended target, before breaking off and

MiG-25P was the first Soviet interceptor offering at least a glimmer of hope for catching one of the SR-71s: eventually, this proved impossible, but pilots of its improved variants, MiG-25PD and MiG-25PDS, kept on trying throughout the 1980s. The 524th Fighter Aviation Regiment – to which this MiG-25PD or PDS belonged – was intentionally deployed at the Letnoezersky AB, in the Arkhangelsk region, with the task of preventing Blackbirds from freely roaming the sky over the Barents Sea and further east.

returning to base. In addition, other types of fighter aircraft such as Su-15 and MiG-23 were also sometimes scrambled from Soviet as well as other Warsaw Pact air bases.

Unlike the MiG-25, which at least had a theoretical chance to engage the SR-71 (though a MiG-25 never fired in anger against one), the other types lacked performance to intercept a Blackbird. That said, a SR-71 could find itself within the reach of those 'red' fighters if at the wrong place and at the wrong time, or if it suffered a technical malfunction negating its speed and altitude advantage. As a matter-of-fact, such a potentially very dangerous situation did indeed happen, but it had a fortunate conclusion. Namely, on 29 June 1987, during a mission over the Baltic, a SR-71 piloted by Lieutenant-Colonel Duane Noll with Lieutenant-Colonel Tom Veltri in the back seat, suffered a failure of its right engine.[22] The Blackbird was forced to descent from its operational altitude down to 25,000 feet (ca 7,600m) as well as to decrease its speed considerably. Finding themselves in a most precarious situation, the Americans decided to head for Swedish airspace. There, they were intercepted by four Swedish Air Force JA-37 Viggen fighters. It needs to be pointed out, that SR-71 missions were not a secret for the Swedes, who nicknamed them 'Baltic Express'.

Even though fighter aircraft in its inventory did not have the performance to intercept a SR-71 flying at its operational altitude and speed, the Swedish Air Force routinely conducted training missions at times when a Blackbird sortie was expected. Thus, when on that particular day misfortune befell one of the SR-71s, two Swedish Viggen fighters were airborne on a training flight and two more, standing by on QRA, were scrambled, with all four promptly dispatched by military air control to intercept the SR-71 which violated Swedish airspace. Soon enough, they sighted the American aircraft, which was obviously in distress and decided to escort it, rendering whatever assistance they could.

According to some sources, the Soviets scrambled up to 20 fighters in an attempt to intercept the Blackbird with Lieutenant-Colonel Veltri allegedly even sighting a MiG-25.[23] Fortunately, the situation had a positive resolution, as the Viggens escorted the SR-71 to friendly airspace without incident and the Blackbird was able to make a safe landing. The events described remained classified for 30 years but once the veil of secrecy was lifted, those involved received the recognition they deserved. Namely, on 28 November 2018 in Stockholm, the capital of Sweden, USAF Major General John Williams presented an Air Medal to each of the former Swedish Air Force Viggen pilots who escorted the Blackbird: Colonel Lars-Eric Blad, Major Roger Möller, Major Krister Sjöberg and Lieutenant Bo Ignell.[24]

The next generation of PVO's fighter-interceptors embodied by the MiG-31, had arguably an even better chance of successfully dealing with the SR-71, if ordered to do so. Among the first units which were equipped with the MiG-31, was the 174 Guards Fighter Aviation Regiment based at Monchegorsk, in the vicinity of Murmansk on the Kola Peninsula.[25] It was an area frequented by 'Blackbirds' flying reconnaissance missions, giving ample opportunity for interception attempts. The latter, as flown by MiG-31 fighter-interceptors of the 174th GIAP and their crews, which would have been similarly conducted in other PVO units, are recounted in the following.[26]

An incoming Blackbird would usually be detected by the radar and ground vectoring station located on the Rybachiy Peninsula. Once the alarm was sounded, crews manned the aircraft which were at the readiness status, meaning they were fuelled, armed, and connected to auxiliary power units. That was not an easy task as the men had to adorn high-altitude flying suits, as well as helmets, and once that was done, encumbered by this outfit, they had to run to the aircraft and climb into the cockpit. The ground crews also had their hands full, amongst other things, they had to remove the R-60 missiles and their launch rails (a fully armed MiG-31 usually had four R-33 under the fuselage and four R-60 under the wings), since they could not be fired at speeds exceeding Mach 1.75 and the MiG would have to fly much faster if it was expected to intercept a Blackbird. Meanwhile, the aircrew was activating and checking the aircraft's systems, starting the engines, and taxiing to the end of the runway. All this was done according to a strict timetable and meant to take up exactly 16 minutes.

When ready, the MiG would stand at the runway's threshold awaiting the order for take-off. Since the SR-71 was flying very fast, very high and keeping its distance outside Soviet airspace, intercepting it required adherence to a rigorous flying regime, meaning that the MiG could not take to the air too late but also, not too early. Having finally taken off, the MiG would climb up to 18500-19500m or even 20,000m, accelerating to Mach 2.35 in the process. Needless to say, this was physically very demanding on the aircrew, especially the pilot who had to work the MiG's controls.

A typical scene from a squadron ready room of one of PVO's regiments of the 1980s. Walls were dominated with posters providing instructions on instrumental landing and schedules for the 1st, 2nd, and 3rd squadron, whilst the speaker's console was decorated by political slogans. (Albert Grandolini Collection)

Intercepting, or more precisely acquiring, the SR-71, was achieved by means of the MiG-31's IRST: the Blackbirds' engines were radiating so much thermal energy that it could be detected from 100km away and sometimes, from an even longer distance. The necessary targeting data was being provided by the IRST, as well as fed by ground control and processed by the aircraft's combat systems. This in turn, signalled the readiness to engage the target with appropriate symbols lighting up on the HUD and the instrument panel, as well as by voice communiques. Because of this, there was no practical reason to activate the MiG's on-board radar, especially as a RC-135 could be lurking somewhere in the distance, just waiting to record the Soviet fighter's electronic emissions.

The on-board radar would only be turned on if the order to engage, that is to actually fire, a R-33 AAM and shoot the SR-71 down, was given, but such a command was never issued. It needs to be pointed out, that in course of an interception, the MiG did not come physically close to the Blackbird, for the latter was, for practical purposes, already intercepted when the fighter's combat systems signalled the readiness to engage it. In addition, the interception was only a brief moment, as high-speed and high-altitude flight could not be maintained for long and the target had to be fired on when it was within engagement parameters. Since a 'live' missile was never launched at a Blackbird, shortly after having intercepted it, the MiG-31 had to break-off. Technicalities aside, there was also a human dimension to all of this, for the aircraft were nothing without the men in their cockpits.

As an example, out of many MiG pilots serving with the 174th GIAP, Major Mikhail Myagkiy flew 14 missions to intercept SR-71s, between 21 August 1984 and 8 January 1987.[27] The sortie which was most memorable for him, took place on 31 January 1986. Having been scrambled and vectored towards the target by ground control, he reached the speed of Mach 2.3 at an altitude of 18,900m (62,007ft). He was still some 60km away from the target when he

An interesting photograph of a MiG-31B intercepted by the Royal Norwegian Air Force while armed only with two old R-40RT air-to-air missiles and none of its usual, and longer-ranged, R-33s. (Albert Grandolini Collection)

DUTY IN THE POLAR DESERT

Whilst PVO operations over the Baltic Sea, the Bering Sea, Kola Peninsula, and the Caucasus are relatively well-known, those over northern Siberia – indeed: within the Polar Circle, in one of most inhospitable climates imaginable – remain almost entirely known. Similar is valid for PVO operations in the Far East of the 1970s and 1980s.[29]

The northernmost permanently occupied air base of the PVO was the Alykel AB, constructed outside the mining town of Norlisk, near the coast of the Kara Sea, in the 1950s. With this facility having the purpose of 'blocking' the closest route between the USA and the Moscow area, it became the home-base of the 57th Guards Fighter Aviation Regiment PVO. Like Amderma AB, Alykel was not connected to the rest of the USSR by roads or railroads, and thus, all the aircraft, supplies, equipment, weapons, and personnel had to be brought there (or flown out) with help of transport aircraft, or the giant Mil Mi-6 helicopters. Foremost, Alykel had no hardened aircraft shelters and – except for a single maintenance hangar – no hangars for the unit's jets: these were always parked outside in the open.

During the winter, ground crews regularly began their working day by digging the aircraft from under 3–4m of snow that fell over the night, and at temperatures below -30°C. The life-line of the 57th Regiment was a fuel and gas pipeline that connected it to the network in the central Soviet Union. Whenever the weather was permitting – foremost during the short summers – it would forward deploy a detachment (or support such a deployment from another unit) to the Greem Bell AB on the Franz Josef Land, only a few hundred kilometres from the North Pole. Conditions on that

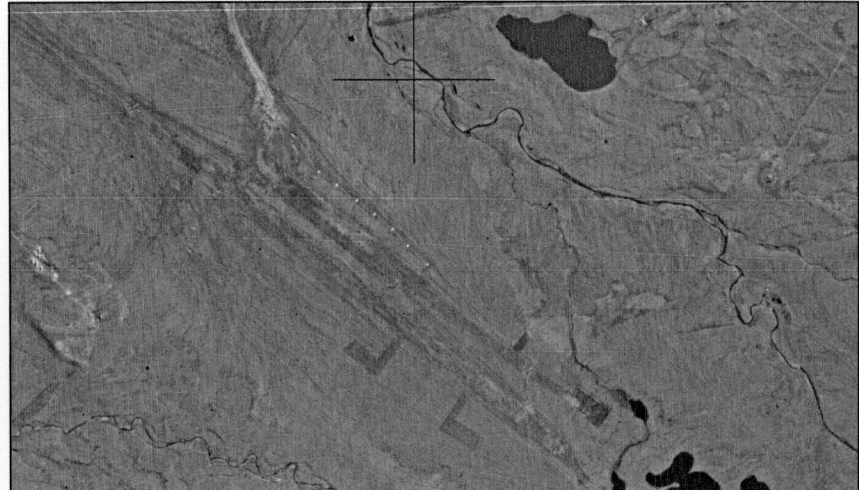

The Tiksi Zapadny AB, as photographed by the KH-9 Hexagon reconnaissance satellite in 1974, when the facility was still little else but 'lines in snow'. In the course of a few summers, this became the air base in the USSR capable of supporting strategic bombers and was the closest to the USA: therefore, it was closely monitored by multiple US intelligence services. (US DoD)

Amderma-based 72nd Guards Fighter Aviation Regiment was one of PVO's units maintaining regular detachments at such forward operating bases like Alykel and Greem Bell. (via Easternorbat.com)

gained a visual (the aircraft's IRST having already acquired it): the SR-71 was about 3,000-4,000m higher, but the Blackbirds' silhouette was just visible at the end of its contrail. Since it was obvious that the US aircraft was on a course which would not bring it into Soviet airspace, the MiG pilot disengaged. From the moment the MiG took to the air, to the point in time its pilot broke off from the target, took only 15 minutes and 40 seconds.

Another noteworthy Blackbird interception was performed by Major Myagkiy on 6 October 1986, the point of interest being that this time, the MiG was sighted by the Americans. On that day, a SR-71, crewed by Lieutenant-Colonel Raymond 'Ed' Yeilding (pilot) and Maj Curt Osterheld, was on a mission outside Soviet airspace in the Murmansk area, flying at an altitude of 75,000ft (about 22,800m) with a speed of Mach 3. The American pilot saw the contrail of an air base were extremely hazardous: even during the summer, the weather was dominated by sudden snowstorms; runway-clearing equipment was not available and pilots frequently had to take-off – and land – not only from icy, but from snow-covered runways. Nevertheless, regular detachments of Su-15s and Tu-128s and, later on, MiG-31s and Su-27s were maintained there as often as possible.

The situation further east was even more primitive: actually, the PVO had next-to-no air bases close to the Arctic coastline until the 1970s, when a large air base was constructed seven kilometres west of the town of Tiksi, on the beach of the Laptev Sea, in the Yakutia region of the Sakha Republic. Originally intended to serve as a staging site for Tupolev Tu-95 strategic bombers underway to strike the USA, the resulting Tiksi Zapadny Air Base was connected to the central USSR by a pipeline, and thus better-equipped and supported. Nevertheless, during the 1970s and 1980s, all the deployments of units of the 14th Independent Air Defence Army, PVO, at this facility remained temporary by nature: only units equipped with long-range surveillance radars were permanently deployed in Tiksi.

In the Far East, the 11th Independent Air Defence Army, PVO, was responsible for protection of a frontline of more than 4,000km. As of the 1980s, this army is known to have comprised four major sub-units:

- 8th Air Defence Corps, HQ in Komsomolsk-na-Amure, comprising four fighter aviation regiments (all amalgamated into the 28th Fighter Aviation Division, in 1980), four anti-aircraft missile brigades, and four radio-technical regiments.
- 23rd Air Defence Corps, HQ in Vladivostok, was the most powerful Soviet air defence asset in the Far East, comprising four fighter aviation regiments, six anti-aircraft missile brigades, one radio-technical brigade and one regiment, and was responsible for the defence of the Vladivostok area, southern Sea of Okhotsk and the island of Sakhalin.
- 6th Air Defence Division, HQ in Petropavlovsk Kamchatskiy, comprised one or two fighter aviation regiments, one anti-aircraft missile brigade, and two radio-technical regiments, and was responsible for air defence of the Bering Sea, Kamchatka Peninsula, and the Sea of Okhotsk.
- 25th Air Defence Division, HQ in Anadyr, comprised three fighter aviation regiments, three anti-aircraft missile regiments, and three radio-technical regiments, and was responsible for air defence of the East Siberian Sea, Chukov Sea, and Bering Sea.[30]

A typical scene from one of the air bases in northern Siberia of the early 1980s: a Su-15TM is waiting for permission to launch for a training sortie, while an Antonov An-14 light transport is arriving with a load of post and supplies. Because of isolation of PVO's air bases, most were heavily dependent on support of transport aircraft for remaining operational. (Albert Grandolini Collection)

A still from a video showing ground crews manhandling a Su-27P at the Greem Bell AB in a snowstorm of the late 1980s. (via Easternorbat.com)

approaching Soviet fighter, which he correctly assumed to be a MiG-31. Lieutenant-Colonel Yeilding could not help but to think of himself and his Soviet counterpart who, obviously unbeknownst to the American, was Major Myagkiy, as of two gallant medieval knights galloping full speed towards each other – except of course, that unlike the MiG, his SR-71 was unarmed. When the MiG reached an altitude which Lieutenant-Colonel Yeilding estimated to be 65,000ft (ca 19,800m) – a remarkably accurate estimation considering what is known about MiG-31 operations – the US airman was able to see it, although it was barely larger than a dot. Shortly thereafter, the MiG dropped off and away whilst the Blackbird continued with its mission.

There were numerous other cases of MiG-31s intercepting SR-71s, not only in the Far North but also in the Far East. In fact, the Soviets believed that their successful interceptions resulted in Blackbird missions being halted. However, there is little in available sources to confirm this. It appears that the winding down, followed eventually by cessation of SR-71 missions along USSR's borders, was a result of the Cold War subsiding, as well as financial pressure (the Blackbird was costly to operate) and the ability to gather most of the data the SR-71 did, by means of satellites rather than any Soviet efforts. That said, PVO's MiG-31s should have been able to stop 'Blackbird' missions by shooting one down, since the R-33 AAM was capable of hitting a target flying at an altitude of up to 28,000m and a speed of Mach 3.5.[28] They chose not to do so because the SR-71 may have skirted Soviet airspace but never flew inside it, whilst downing one in international airspace would have been an incident raising Cold War tensions to a level nobody (especially not the Soviets in the 1980s) desired.

Whilst the Soviets were confident that a MiG-31 could bring a SR-71 down if ordered to do so, Blackbird crewmen, like the aforementioned Lieutenant-Colonel Yeilding and Major Osterheld, were of the opinion that due to a number of factors, the odds of them being hit by an AAM (should one have been launched), were low. In retrospect, it is arguably better that this remained a purely theoretical question and nobody had to practically confirm it either way.

5
BORDERLINE BEHAVIOUR

As spectacular as they were, high-altitude interceptions of the sort described in the previous chapter, were nothing but a small fraction of PVO's daily work. For the most part, it was called to intercept countless military and civilian aircraft which flew in the vicinity the USSR's borders or actually violated Soviet airspace. Obviously, intercepting foreign reconnaissance aircraft was of greatest importance. Such ranged from small fighter-reconnaissance types up to large four-engine ones such as the American RC-135s. Reconnaissance missions were flown by American, other Western and allied aircraft and also, by aircraft belonging to ostensibly, neutral countries. For example, Finland's Ilmavoimat Il-28Rs were observed flying along the USSR's border but over Finnish territory. Considering that the Ilyushins were of Soviet manufacture, this clearly demonstrates that 'trolling' is not a recent phenomenon. Similarly, foreign maritime patrol/reconnaissance aircraft also frequently ventured close to the Soviet airspace. Additionally, aircraft of various other types flew in the vicinity of the Soviet border or actually crossed over it into USSR's airspace, resulting in a number of incidents, including a few shootdowns. In some areas, incidents were more frequent than in others. For example, the Soviet 'soft underbelly' in the south along the borders with Iran (up to the 1979 Islamic revolution a close ally of the United States) and Turkey, was an area where such were relatively frequent.

Persian Prancing
During most of the Cold War, bar little more than its final decade, Iran was a close ally of the United States. This meant that its border with the Soviet Union was literally, a frontline of the Cold War which in turn, meant that the likelihood of incidents, including aerial ones, taking place there was particularly high. Indeed, through the 1960s and 70s, several Iranian aircraft, as well as two helicopters, were either shot down or forced to land by the Soviets.

The first instance of such is somewhat mysterious, in that it lacks confirmation from the Soviet side. Namely, on 4 August 1961, a Douglas DC-4 of the Iranian Airways, registered as EP-ADK, was flying back to Tehran after a cargo flight to Beirut. The aircraft flew into Soviet airspace apparently due to a navigational error. It was attacked, by presumably Soviet fighter aircraft, and as a result, its number 1 engine caught fire, prompting the crew to extinguish it by feathering the propeller. Shortly thereafter, the number 4 engine had to be shut-down too and the propeller feathered because of a number 4 fuel tank low fuel indication. This situation forced the crew to make an emergency, wheels-up landing on the southwest coast of the Caspian Sea. Interestingly, as far as it could be ascertained from available sources, there is no Soviet claim corresponding with this incident. However, an even more serious incident took place two years later.[1]

Dramatic events unfolded in the early morning hours of 20 November 1963 as an aircraft crossed into Soviet airspace from Iran. Two MiG-19s of the 152nd Fighter Aviation Regiment were scrambled from Ak-Tepe and two additional ones from the 156th Fighter Aviation Regiment from Mary-2.[2] The former failed to intercept the intruder but the latter had more luck. Having intercepted the airspace violator which was identified as a 'military L-26' (even though a civilian registration was noted), Captain Pavlovski twice signalled it to follow. The foreign aircraft did not comply and turned towards Iran, whereupon ground control instructed the Soviet pilot to prevent it from crossing the border. Captain Pavlovski opened up with his MiG's 23mm gun, crippling the targeted aircraft but it went down already inside Iranian territory. Supposedly, the Soviets saw three men climbing out of its wreck.

Tehran protested the incident accusing the Soviet Union of shooting-down a civilian aircraft inside Iranian airspace, indeed, 'near Mashhad'. However, the Soviets had different ideas as to the event's nature. First of all, the incident took place when a Soviet delegation, headed by Leonid Brezhnev (he was soon to become the Soviet Union's leader), was on an official visit to Iran. Moreover, not only did the incident coincide with the aforementioned visit but in addition, the KGB claimed to have uncovered that Iranian intelligence operatives, as well as a CIA agent, were aboard the aircraft. Unsurprisingly, the Soviets concluded that it was a deliberate provocation.[3]

Such assertions seem to lack substance in view of available information: the aircraft was an Iranian civilian Rockwell International Aero Commander 500 (construction number 500-846 98, former US registration N8487C, Iranian registration EP-AEL): three men were on-board, including two cameramen and the pilot – and only the latter survived.[4]

Another Iranian Aero Commander was intercepted inside Soviet airspace a year later in 1964, by a MiG of the 156th Fighter Aviation Regiment, flown by Captain Pechenkin.[5] On this occasion, the airspace violator did not attempt to escape but landed in Soviet territory as instructed. An inspection on the ground revealed that the aircraft was equipped for photomapping. Officially, the Iranians were engaged in a civilian project to map the border areas but the Soviets were convinced that they just had foiled an attempt at aerial espionage.[6] Yet another Iranian aerial intruder, this time a De Havilland Canada L-20 Beaver, was forced to land on 28 June 1967 by one of 156th Fighter Aviation Regiment's MiGs with Captain Stepanov at the controls.[7] The aircraft and its crew were released to Iran a few days later.

On 6 October 1970, an Aero Commander 500B involved in the operation of mapping on behalf of the Cartography Department of the Iranian government (construction number 500B-1631-217, registration EP-ALP), flew about 1,200m deep into the Soviet airspace. It was intercepted by a MiG-21, which opened fire without any kind of warning. Although the tail and the right engine of the Iranian aircraft were badly damaged, the pilot managed to fly it back to the Iranian airspace and for 80km south of it, before making an emergency landing about 27km north-west of Mianeh: both occupants came away unharmed. Yet another incident involving an Iranian Aero Commander was reported by the Soviets on 21 June 1973: the aircraft was intercepted by a Su-15 from the 976th Fighter Aviation Regiment and forced to land in the USSR, but subsequently left to return to its homeland.[8]

However, PVO fighters were not always successful. A particularly embarrassing incident taking place on 8 October 1977 when an Iranian light aircraft, the pilot of which apparently lost his bearings, entered Soviet airspace. The aircraft landed on the Nebig-Dag airfield where the Su-15 equipped 354th Fighter Aviation Regiment was based (*sic!*).[9] Once the Iranian pilot realised where he is, he promptly took off and flew back to Iran. The Soviets failed to intercept the incoming airspace violator, were unable to arrest the aircraft on the ground and did not prevent it from flying back to Iran either. To say that PVO's command was unhappy when it learned of this failure, is an understatement. Thus, harsh punitive measures followed: a number of officers were penalised, and the unit was disbanded.

MiG vs Chinooks

Due to its seriousness and the number of resulting casualties, an aerial incident along the Soviet-Iranian border, stands out in particular. Namely, at 06:21 on the morning of 21 June 1978, a Soviet radar site near the village of Bagir in Turkmenistan, detected four slow-moving contacts coming from the direction of Iran, violated Soviet airspace ca 15-20km deep in the vicinity of the town of Dushak, in the Soviet Turkmenistan..[10] Five minutes later, these contacts were also detected by the radar site at the Ak-Tepe airfield where the 152nd Fighter Aviation Regiment – meanwhile re-equipped with MiG-23Ms – was based. Its deputy commander Lieutenant Colonel Miloslavsky, ordered a MiG into the air, piloted by Captain Demyanov. However, he sighted only a single helicopter and misidentified it as a friendly one. Since Captain Demyanov's communications with ground control revealed his uncertainty, he was ordered to return back to base.

Another MiG-23 with Captain Valery Shkinder at the controls, was scrambled at 06:52 and guided by ground control towards the airspace violators. Captain Shkinder approached the four contacts and correctly identified them as Boeing CH-47 Chinook helicopters whereupon he received an order to attack them. At the time, the Iranian helicopters were flying on a north-western course along the Garagum Canal. However, once the Iranians became aware of the Soviet fighter, they changed to a southwestern heading trying to reach Iranian airspace. Not all made it across the border for meanwhile, Captain Shkinder manoeuvred his MiG behind the rearmost helicopter and launched two R-60 AAMs against it. Both missiles struck the intended target shooting the Chinook down. It crashed in the vicinity of the village Gjaurs, killing all eight men who were aboard.

Captain Shkinder reported the destruction of the first target to ground control and was ordered to engage the second one. Accordingly, he made two passes, firing the MiG's GSh-23 cannon at another Iranian helicopter. In all, 72 rounds of 23mm calibre were expended and as a result, the CH-47 took hits to one of its engines. The Iranian pilot was able to land the damaged helicopter with all four men who were aboard it, apparently surviving unscathed. However, the helicopter touched down in the vicinity of the Soviet border post at Gjaurs and all the Iranians were detained by Soviet border guards. Meanwhile, the two remaining CH-47s managed to fly away back into Iranian airspace.

It was obviously a very serious incident yet neither the Soviets nor the Iranians had the desire to escalate. Thus, the Soviets permitted the damaged helicopter to be repaired and flown back to Iran: the crew were also released. As for Captain Shkinder, he was

The MiG-23M brought the much-required flexibility to the PVO: thanks to its look-down/shoot-down radar and fire-control system, and armament including R-60 short-range air-to-air missiles ('AA-8 Aphid') it could effectively operate against low and slow-flying targets such as light aircraft and helicopters. Captain Shkinder of the 152nd IAP demonstrated all of this in dramatic fashion on 21 June 1978. (Albert Grandolini Collection)

recommended for the Order of the Red Banner, but the decoration was not awarded for political reasons (simply put, the Kremlin wanted to close the matter without drawing more attention to it): instead, he was reposted to another unit. As a side note: that was the first case of a PVO fighter-interceptor opening fire at helicopters and as a result, the incident was used to draw up official instruction on how to engage this type of target.

After the Islamic revolution of 1979, the number of aerial incidents along the Soviet-Iranian border decreased considerably. Being known for their anti-Americanism, the new rulers in Teheran would not play Cold War games on USA's behalf and anyway, they had to worry about the war with Iraq. That said, on 23 December 1979, an Iranian Cessna was forced to land in Soviet territory.[11] A Su-15 flown by Captain Kondakov failed to spot the airspace violator but forced it to land, nonetheless. When the Iranians saw an armed fighter-interceptor overhead, they decided not to take any chances and landed on a road.[12] Other than that, during the 1980s, instances of Iranian aircraft (both fixed-wing as well as helicopters) flying along the Soviet border were observed, yet it seems airspace violations were few and far between, with just one noted in 1987.[13] Apparently, over the next few years, it was only on 2 September 1990 when an Iranian military aircraft was forced to land in Nakhichevan in the Soviet Azerbaijan, that another violation took place..[14]

As a general rule, the Soviets were not interested in ratcheting up tensions with Iran, be it Imperial or Islamic, and because of this, in virtually every case, the pilots (and other persons who were aboard), as well as the aircraft, were sooner or later released. It appears most incidents described thus far, were indeed the results of navigational errors, the loss of orientation and such. However, there was also aerial reconnaissance activity undertaken from Iranian territory resulting in a serious incident which warrants a separate description. In addition, there were a few other incidents both involving manned aircraft as well as aerostats in the vicinity of the Soviet-Iranian border. All this will be related in the sub-chapters and chapters to follow.

Turkish Troubles
Another Cold War frontline state, sharing a border with the USSR, was Turkey which was and is, not only an ally of the United States but also a member of NATO. Unsurprisingly, there were several Cold War aerial incidents with the Soviets involving both American as well as Turkish aircraft.

A particularly interesting case took place on 21 October 1970.[15] It so happened that a Beechcraft U-8 Seminole of the US Air Force (US-FY-serial 58-3085), piloted by Major James P. Russell jr. with Major General Edward Scherrer (head of the United States military mission in Turkey) and Brigadier General Claude Munro McCurry (head of the Army section of the mission) as well as Colonel Kevat Deneli (an officer of the Turkish Army) aboard, was en route from the Turkish town Erzurum to Kars, also in Turkey.[16] Due to a combination of cloud cover limiting visibility, as well as strong winds, the aircraft diverged from its intended course apparently, without the pilot realising. Unfortunately, the aircraft had strayed over Soviet Armenia and was intercepted by Su-15s of the 166th IAP and forced to land in Leninakan (nowadays Giumiri).[17] It took lengthy negotiations to release the US aircraft and all four men were only allowed to fly away on 10 November.

In other cases, Turkish aircraft – mostly fighter-reconnaissance types – violated the Soviet airspace on a number of occasions but, due to various circumstances, neither the manned interceptors nor SAMs of the PVO, managed to counter them successfully. Almost exactly 39 years later, Turkey caused much uproar – indeed: a scare that might have caused the Third World War – when General Dynamics F-16C Fighting Falcon fighter-bombers of its air force (Türk Havva Kuvvetleri, THK), intercepted and shot down a Sukhoi Su-24M fighter-bomber of the Russian Air & Space Force after it made 'just a shallow' penetration of the Turkish airspace: what is unknown – indeed: largely forgotten even in Turkey – is that on 24 August 1976, the Soviets shot down a THK jet for doing much less.

At 09:12hrs of that morning, two Northrop RF-5A Freedom Fighters of the 184 Filo (squadron) launched from the 8th Main Jet Base. Piloted by 1st Lieutenant Sahir Beceren and 1st Lieutenant Hikmet Bogatir, they were to run a photo-reconnaissance mission of the towns of Siirt, Mus and Agri. After passing Mus, both aircraft turned for Agri before (according to contemporary press reports) deviating from the planned route – probably due to a minor disorientation or some sort of technical malfunction – and flew into Soviet airspace. A subsequent investigation by the THK revealed 'just a shallow' penetration: mere 1,000–1,500m. Nevertheless, the Soviets took it extremely seriously and fired four SAMs at the two RF-5As. At around 09:55hrs, one of these hit the jet flown by Bogatir, as this was underway at an altitude of about 25,000ft (7,600m). The stricken Freedom Fighter crashed about 7km south of the town of Igdir, but 1st Lieutenant Bogatir managed to eject safely and was assisted by the locals once on the ground. Turkish sources never mentioned the type of the SAM that shot down this RF-5A, but the Russians quoted it as S-125 ('SA-3 Goa').[18]

Maritime Mischief
Intercepting foreign reconnaissance aircraft was a routine task for Soviet fighters. Just how frequent PVO fighter aviation's interceptions were, can be illustrated by the example of the 865th IAP, which is known to have intercepted Boeing RC-135s strategic reconnaissance aircraft of the USAF 33 times in 1981 alone, and another 128 times in 1985. However, unlike in the 1950s – bar possibly in course of one controversial incident described in the sub-chapter Non-events? – not a single such reconnaissance aircraft was shot down between 1961 and 1991. That said, they were occasionally harassed, in particular when flown over the sea. For example, on 24 September 1962, a Boeing RB-47H (53-4287) of the USAF had a very close encounter with Soviet interceptors over the Baltic Sea east of Gotland Island. The A/C John Drost took a photograph of one of the aircraft, a MiG-19, which was almost close enough to touch.[19]

Supposedly it got even more serious when another RB-47H was attacked by a MiG-19 in 1963.[20] However, since any details are unknown – for example the date and location of the incident – one can be sceptical if it ever happened at all. On 17 November 1970, when in the vicinity of Vaygach Island in the Pechora Sea, two MiG-17s fired warning shots at an KC-135T (55-3121), apparently trying to compel it to fly away.21 However, since the US aircraft was over international waters, it continued with its mission and the Soviets did not escalate any further: the MiGs escorted the KC-135 but abstained from firing more shots. Interestingly, as far as the author was able to ascertain, the Soviets (nowadays Russians) made hardly any mention of those incidents. This was possibly due to the fact that they did not result in loss of life or injury and there was no material damage either. However, there was a different story with a few serious incidents involving maritime reconnaissance aircraft.

Neutral Swedes
Despite being ostensibly neutral during the Cold War, it was common knowledge that Sweden had a 'somewhat Western tilt'. Moreover, due

By the time of the 'dogfight' on 7 July 1985 between the Swedish SH.37 Viggen and the pair of Su-15TMs of the 54th Guards Fighter Aviation Regiment, all of the latter were upgraded through the installation of R-60 short-range missiles on inboard underwing pylons. Most received a coat of camouflage colours – as visible on this example from an unknown unit. (Albert Grandolini Collection)

chase but the Swedish jet's RWR showed that it had locked-on the SH.37, with an apparent intent to avenge his comrade. Fortunately, a pair of JA.37 Viggen interceptors, airborne in the vicinity of Gotland Island, flew to Captain Larsson's assistance, whereupon the Soviet pilot disengaged. The Su-15 flew back to the crash site, circled over it for some time and finally headed for its base.

Meanwhile, Captain Larsson returned to base and the Swedes had to deal with a situation that could potentially, have had very serious repercussions. Fortunately, the commander of the Swedish Air Force, Lieutenant General Sven-Olof Olson, officially expressed his regret about the incident and tension resulting from it disseminated quickly. In fact, it never really built-up as the USSR did not even file a formal note of protest.

The Neptune Incident

The incident described previously, remained an exception as the tracking of submarines, shadowing of surface warships, observing Soviet naval activity et cetera, was usually performed by American, as well as other western and allied maritime patrol (reconnaissance) aircraft. Such were, for the most part, large and slow, two or four-

to ties with the US and NATO, which at that time, were not publicly known, the Swedes were engaged in reconnaissance activity, not only for their own benefit but also for the benefit of other western nations. As a result, encounters between Swedish and Soviet aircraft over the Baltic, were not uncommon. Usually, there was nothing dramatic or even noteworthy about them however, the one on 7 July 1985 turned deadly.

On that day, a SH.37 Viggen fighter-reconnaissance aircraft from the Bråvalla flygflottilj F13, Swedish Air Force, (serial '03 Red'), took off from its base at Norrkoping with Captain Göran Larsson at the controls.22 It was tasked with observing a group of Warsaw Pact warships exercising in international waters off the coast of the Soviet Latvia. As the Swedish aircraft approached the vessels in question, two Su-15TM fighter-interceptors of the 54th GIAP, based at Vainode AB, appeared. Each was armed with two R-98 and two R-60 AAMs, both wore recently-introduced camouflage colours, and Captain Zhigulyov's aircraft is known to have worn the bort '36 Yellow'.

The ships were scattered over a wide area and one of the Sukhois come too close to the Swedish aircraft for comfort (the other kept his distance). Under the circumstances, Captain Larsson returned to base, topped off his fuel tanks and became airborne, returning to the naval exercise area to try his luck a second time. Flying at low level and maintaining radio as well as radar silence, the Swedish pilot hoped to avoid detection and the unwelcomed company of Soviet fighters. However, not only did the same pair of Su-15s appear again, but '36 Yellow' began tailing the Viggen aggressively.

Wanting to shake off the pursuer, Captain Larsson performed a series of tight manoeuvres at low level and high-speed. Being less manoeuvrable than a Viggen, the Su-15 could not follow, but nonetheless, Captain Zhigulyov gave it a try – and paid for this with his life. Suddenly, in his rear-view mirror, Captain Larsson saw a big splash on the surface of the sea, followed by an explosion. The Soviet fighter crashed and there was no ejection meaning that Captain Zhigulyov was killed.

Playing a Cold War game had unexpectedly, turned deadly serious, in the most literal sense. In view of this sudden and dramatic development, Captain Larsson decided to head home as fast as possible. Flying low, he lit the afterburner and went supersonic but he was not out of trouble just yet. The surviving Su-15 not only gave

Tsymbal's Su-27P as seen while approaching the Norwegian P-3B on 13 September 1987, with its big air brake open to slow down to the docile speed of the Orion. (Royal Norwegian Air Force)

A close-up photograph of Senior Lieutenant Vasiliy Tsymbal, as he took a position right next to the left wingtip of the Norwegian Orion, about a minute before the collision. (Royal Norwegian Air Force)

Norwegian P-3B serial 602, as seen with its engine Number 4 stopped and damaged propeller tips, on return to the national airspace. (Royal Norwegian Air Force)

engine propeller-driven types. Insofar these aircraft operated in relative proximity to the Soviet coast, they were frequently intercepted and subsequently shadowed (escorted) by PVO fighters. Occasionally, Soviet fighters harassed them trying to disrupt their mission.

Fortunately, this seldom degenerated into a serious incident but there were exceptions: in one instance, a Japanese aircraft was almost shot down, whilst on another occasion, a Norwegian aircraft sustained damage, together with the Soviet fighter which intercepted it. Concerning the first incident, it started on 2 April 1976 when a Su-15 fighter-interceptor, with Senior Lieutenant Strizhak of the 777 IAP, was scrambled from Sokol airbase to intercept a USAF RC-135 reconnaissance aircraft that had approached within 100km of the Sakhalin Island.[23] After take-off, the Su-15 was redirected by ground control to intercept a JMSDF P-2 Neptune. The Neptune was flying over the Sea of Japan at ca 2000m altitude near the southern tip of Sakhalin. Approaching within 4–6km of the P-2, the Su-15 followed the former on a parallel course. It would have been another routine interception however, things turned very dramatic when Senior Lieutenant Strizhak inadvertently fired an AAM at the Neptune. Fortunately, he had the presence of mind to swiftly break-off the engagement sequence causing the missile to harmlessly pass off the Japanese aircraft's right wing and self-destruct.

Over a decade later, another incident resulted with a Soviet fighter and a western maritime patrol aircraft having an encounter too close for comfort. On 13 September 1987, a Su-27 '36 Red' of the 941st IAP was scrambled from Kilpyavr.[24] Its pilot, Senior Lieutenant Vasiliy Tsymbal, was vectored to intercept a western aircraft which turned out to be a Norwegian Air Force (*Luftforsvaret*) P-3B Orion, serial number 602 of the No 333 Squadron from Andöya AB, flown by Lieutenant Jan Salvesen and a nine man crew.

The Soviet pilot not only intercepted the Norwegian aircraft but, by manoeuvring close to it, tried to disrupt its mission. Two such close encounters took place with the Su-27 accelerating away, only to eventually come back. Appearing for the third time, the Soviet pilot flew under the starboard (right) wing of the Orion, obviously intending to pull up in front of the Norwegian aircraft.

However, while doing so, the right vertical stabiliser of his Sukhoi contacted the propeller of the Orion's Number 4 engine. The dielectric cap topping the fin was shattered but the propeller also cut into metal. As a result, a 11cm piece of the propeller broke off and struck the P-3's fuselage puncturing it and causing decompression: moreover, violent vibrations caused by the damaged propeller, forced a shut-down of the number 4 engine. Fortunately, no-one was physically injured and both aircraft separated, each returning to its base safely. After the incident, Senior Lieutenant Tsymbal was reposted to another unit whilst the Sukhoi involved had its number changed first to '38 Red' and subsequently to '31 Blue': additionally, it received a 'kill marking' in from of a blue P-3 silhouette.

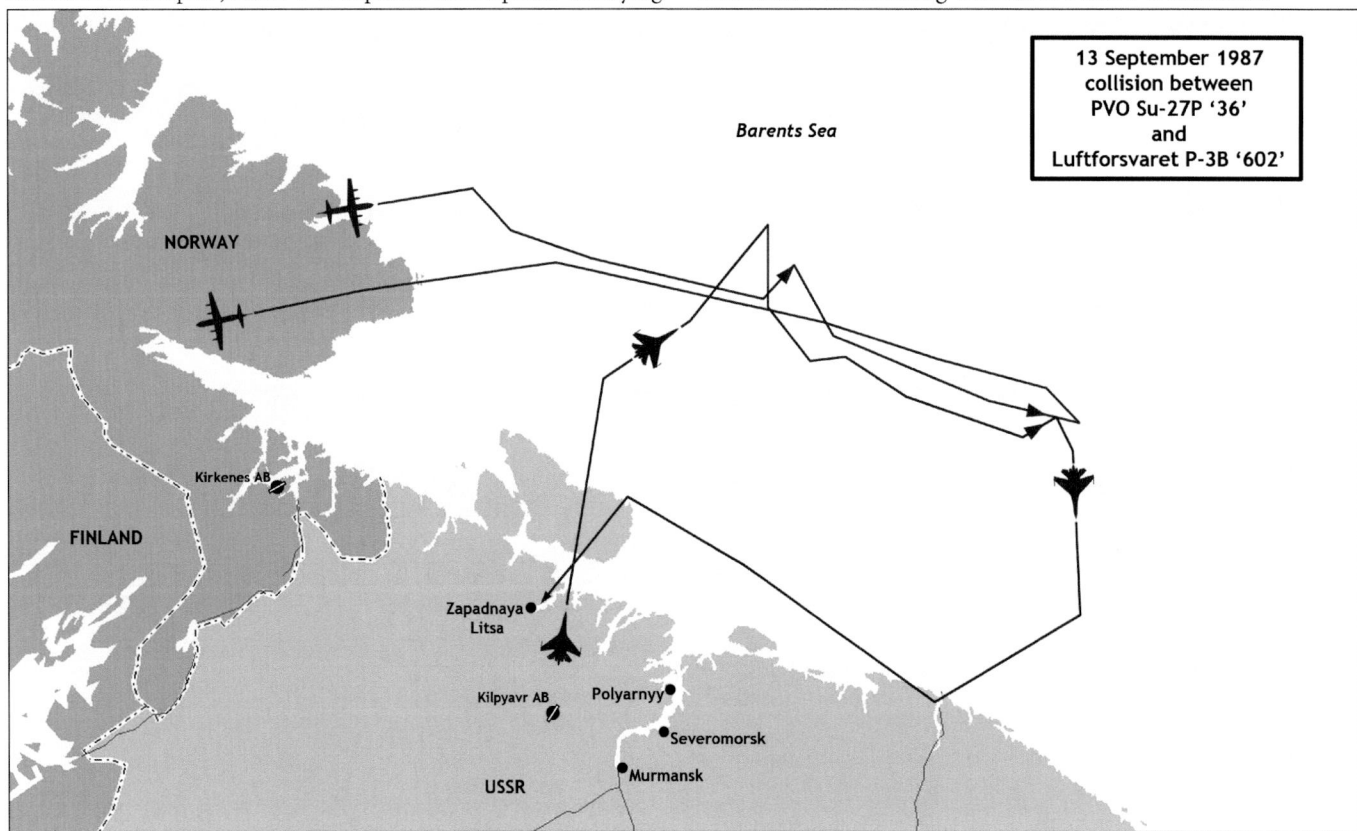

A map of the encounter between the Norwegian P-3B Orion and the PVO's Su-27P. (Map by Tom Cooper)

Another of PVO's Su-27Ps encountered by the Royal Norwegian Air Force during September 1987. (US DoD)

The incident was noteworthy, not only because of its obviously serious nature, but also because at that time, the Su-27 had only relatively recently entered service and was considered a Soviet 'super fighter' in the West. Little solid information was available about it until, in the course of the events described, the Norwegian airmen were able to take detailed photographs of the Su-27. The said photos were soon released for publication and appeared in a number of outlets, for example the 26 September 1987 issue of the Flight International magazine.

Red on Red

Thus far, incidents involving foreign aircraft have been described. However, the PVO's efforts to deal with real or perceived airspace violators, also resulted in a number of fratricidal incidents, which, whilst infrequent, still took place on a few occasions. There is one known instance when PVO SAMs shot down a PVO fighter. Namely, in the mid-1960s, the 2nd *divizion* (battalion) of the 441st rocket (missile) Anti-Aircraft Missile Regiment, by means of a S-75 ('SA-2') surface-to-air missile, shot down a Yak-25 fighter-interceptor in the vicinity of Gornostay bay near Vladivostok.[25] Fortunately, the Yak's two-man crew managed to eject. As was subsequently determined, the aircraft's IFF transponder failed and it seems the PVO could not allow an unidentified aircraft to loiter around such a sensitive area as Vladivostok with its many military facilities.

Additionally, there were a number of close calls. On at least one occasion, two Soviet fighters were scrambled to intercept a supposed airspace violator, only to be vectored by GCI, one against the other. Fortunately, the mistake was realised before any shots were fired.[26] Apart from fighters, other aircraft types were involved too: on 24 March 1983, a Tu-22 bomber flew into Iran due to a navigational error and caused alarm in the PVO when it returned to Soviet airspace: fortunately for its crew, it was identified positively on time and did not result in a fratricide.[27] In addition, a Soviet passenger aircraft on a domestic flight was intercepted by a PVO fighter but everything ended with just a scare.[28]

In the meantime, training, weapon tests and live firing exercises, brought about a few more cases of fratricide. The crash of a Tu-104 passenger aircraft on 30 June 1962, some 28km east of Krasnoyarsk airport resulting in 82 fatalities (all who were aboard), may have been caused by a SAM fired in course of a live firing exercise but this was never officially confirmed.[29] In the early 1970s, a Su-11 of the 393rd GIAP, while practising interception techniques, fired a 'live' R-8 AAM against a Su-9, shooting it down.[30] Luckily, the pilot of the latter managed to eject and save his life. Apparently, an aircraft standing by on alert status (meaning fuelled and armed with live ordnance), was used in the exercise. Its pilot realised too late that having locked-on and pressed the trigger, he launched a

During the early 1980s, numerous PVO regiments were re-equipped with MiG-23ML interceptors: smaller but as fast as older types, they proved much more flexible, and were available in large numbers. On 8 June 1985, weapons testing of their new R-24 missiles went wrong, resulting in the tragic loss of an An-26 tasked with monitoring the trial. (Albert Grandolini Collection)

real AAM which physically destroyed the target before he could do anything about it.

A similar incident involved two Su-9s but on that occasion, the pilot had the presence of mind to break-off the engagement sequence before the already launched AAMs reached the target.[31] In another instance, on 8 June 1985, while testing new work modes of its on-board radar at Aktyubinsk, a MiG-23ML fighter flown by Colonel Anatoly Sokovykh, launched an R-24T infra-red homing, medium-range air-to-air missile against a practice target.[32] Due to a fatal mistake of the flight controller, another aircraft, an An-26 equipped as a flying laboratory, was allowed to fly in the same area at the same time. Instead, on the practice target, the MiG's radar locked onto the Antonov, guiding the missile against it. Even though it was a practice AAM with an inert warhead, its impact still chopped off one of the An-26's wings. As a result, the aircraft plunged to earth killing its eight man crew when it hit the ground. Yet another fratricidal incident took place during a live firing exercise on 18 August 1989 at Sary Shagan, when one MiG-31B of the 174th GIAP mistakenly shot down another MiG-31B of the same unit by a single R-33.[33] The crew of the downed interceptor, including pilot Major Kudravcev and navigator Captain Pogorelov ejected, but the latter succumbed to his injuries.

Non-events?

Last but not least, there were a number of possible incidents which may have resulted in the shooting-down of foreign aircraft by the Soviets, but which are far from being conclusively confirmed. Among such, the loss stands out of the RB-57 'Canberra' (63-13287) of the 7407 Support Squadron, based in Germany but on temporary detachment in Turkey, on 14 December 1965 over the Black Sea. It was a tragic incident resulting in the RB-57's both crewmembers: Captain Lester L. Lackey and Lieutenant Robert A. Yates, losing their lives.

At the time of the 'Canberra's' loss, Soviet fighters were airborne but not in the immediate vicinity.[34] In a case study of the incident, it was stated that wreckage of the aircraft which was recovered did not show any obvious signs of attack by hostile fighters.[35] Likewise, in another study which analysed shoot downs of US aircraft in a case-by-case manner, the possibility that the RB-57 was shot down, whilst mentioned, was dismissed.[36]

It appears the US-Russia Joint Commission on POWs/MIAs did not conduct an in-depth investigation of this matter which it would have done, as with other similar cases, if there were serious indications that the RB-57 was shot down by the Soviets. Having said this, one cannot reject the incident to the 'case closed' category. There is an anonymous, yet credible, testimony that recovered RB-57 wreckage bore signs of pre-crash damage, most likely resulting from a SAM hit.[37] In addition, both crewmen were awarded Purple Heart medals which are bestowed upon only those US service members who were wounded or killed by hostile fire. Thus, whilst still not definitely confirmed, it is very possible that the 14 December 1965 RB-57 loss was indeed, the result of Soviet hostile action.

There were also claims that a reconnaissance aircraft (other than the aircraft described in the sub-chapter Persian Prancing and the November 1973 Phantom loss which will be related in the chapters to follow) operating from Iran, in course of US-Iranian reconnaissance operations, was shot down by the Soviets. In particular, there were journalistic assertions that in 1967, an ERB-47H was shot down over the Caspian Sea. However, this lacks confirmation and is more likely than not, nothing but a sensationalist claim.[38]

Finally the Sino-Soviet split and the conflict it brought about, may have also resulted in the downing of two Chinese J-7 fighters (a Chinese 'clone' of the Soviet MiG-21 ASCC/NATO-codename 'Fishbed') in the time period 1975-76: one by Soviet MiG-23 fighters and the other ambushed by ground troops armed with Strela MANPADS.[39] Having noted those possible, yet unconfirmed incidents, it must be stated that the final word in regard to the history of Cold War aerial incidents, has not been written yet.[40]

Apart from the ones described thus far, a number of other incidents took place as well, which due to their nature, circumstances, types of aircraft involved et cetera merit a separate description.

6
BALLOON BUSTING

Arguably, the vast majority of professional military aviation researchers, as well as amateur enthusiasts, would most likely associate balloon busting with the First World War. Yet if numbers are anything to go, by it was also an important part of the Cold War. As far as it could be established, no totals were ever published in Russian or other post-Soviet sources, but judging by available data, combating aerostats was an important task of PVO's fighter aviation. Namely, between 1956 and 1977, no less than 4,112 aerostats were detected by the Soviet PVO, out of which, 793 were shot down.[1]

In the time 11 August – 14 September 1975, as many as 11 foreign aerostats floated into Soviet airspace, usually drifting at an altitude of ca 11–14000m with a speed of up to 200km/h.[2] In order to deal with each, sometimes just one and at times, up to 16 fighter-interceptors were scrambled of the types: MiG-19, MiG-21, Yak-28, Su-15 and even Tu-128. As a result of their attacks, eight balloons were shot down, two had their gondolas shot away and one drifted out of Soviet airspace unscathed. At first glance, it could appear that an aerostat was an easy target, ordnance expenditure tells a different story, for in order to shoot one down, it took on average 1.4 AAMs, 26 unguided rockets and 112 cannon shells – those balloons were apparently, tough nuts to crack!

Fiddling Around

As stated, a comprehensive listing of all balloon shootdowns by the PVO seems to be lacking. However, information regarding numerous incidents involving aerostats can be extracted from available data. As an example, on 3 April 1966, a Yak-25 of the 146th Guards Fighter Aviation Regiment, shot down a balloon; similarly, a Yak-25 of the same unit crewed by Major Volkv (pilot) and Captain Vakhrushev (radar operator) shot down another one in June 1968.[3]

Bringing a balloon down sometimes required the coordinated action of several aircraft. On 29 October 1967, it took no less than three of 865 IAP's fighters to shoot one down: a MiG-17 flown by Captain Avtushko, as well as two Su-9s, piloted by Major Lysyura and Captain Plakhotnik.[4] In another instance, an aerostat was shot down on 14 December 1969 by a Yak-28 of the 174th GIAP flown by Captain Chernega.

At times it was a quite challenging task: on 3 July 1973 Captain Blinov of the 865th IAP, flying a Su-9, shot down an aerostat drifting at an altitude of 21,000m (68,898ft) which was at the very edge of the Sukhoi's capabilities. For this feat he was awarded with a personalised hunting rifle.[5] A few pilots could boast more than one balloon 'kill': while serving with the 562nd IAP Captain Sninshikov brought a balloon down on 19 August and 5 September 1975.[6] He was flying a MiG-19 on the former and a Yak-28, on the latter date.

A noteworthy incident took place in April 1984 (unfortunately the exact date remains unknown) when several MiGs, including a MiG-25 of the 82nd IAP, were scrambled to intercept an aerostat which drifted into Soviet airspace from Iran.[7] The balloon was floating at an altitude of ca 25,000m when it was struck by a R-40 air-to-air missile fired by the MiG-25 (possibly the highest

A rare air-to-air photograph, showing a Tu-128 armed with two R-4 air-to-air missiles underway on a combat air patrol, at a very high-altitude. (Albert Grandolini Collection)

A Yak-28P of the 82nd IAP, armed with an R-3S (visible on the outboard underwing pylon) and an R-8R, (inboard underwing pylon), standing QRA at the Nasosnaya AB, outside Baku, in the early 1970s. The unit was one of most successful of the PVO in the discipline of intercepting so-called 'spy blimps'. (via Easternorbat.com)

A MiG-25PD – here seen armed with a single R-40U training round on the inboard underwing pylon – of the 82nd Fighter Aviation Regiment, as seen in the late 1970s, before all the jets of that unit were upgraded to the MiG-25PD-standard. (via Easternorbat.com)

ever AAM shot fired in anger and not in training, during testing et cetera). Even though the missile failed to explode, it tore through the aerostat's envelope and as a result, it started to lose altitude. When it was down to ca 16,000m it was finished-off with gunfire by MiG-23s.

For many PVO fighter-interceptor units equipped with types like Yak-25 and 28, the Su-9, and others, aerostats were the only real-life targets (not practice ones) at which they had an opportunity to fire in anger. Among them was the Tu-128, the largest interceptor entering frontline service anywhere in the world. Thus, the 'Fiddler', this being its ASCC/NATO-codename, only got to fiddle around with balloons. During the late 1970s, one of 518 AP crews composed of Major V. Sirotkin (pilot) and E. Shchetkin (radar operator/navigator), shot down two aerostats: one in the vicinity of the Kolguev Island and the other in the area of the town Naryan-Mar.[8] In order to bring the latter down, they had to fire four AAMs, which is the Tu-128's entire combat load. Another foreign balloon was destroyed over the Yakutsk region on 19 July 1985 by one of 350, Fighter Aviation Regiment's Tupolevs crewed by Major N. Savoteev (pilot) and Major V. Shirochenko (radar operator/navigator).

Notably, not all attempts to shoot balloons down were successful. For two days 31 March – 1 April 1970, the Omsk based 64 AP would send up to three Tu-128s, in an effort to bring down an aerostat floating at ca 21000m, but all attempts proved fruitless. This failure resulted in a conference being organised by PVO fighter aviation's command with the aircrews involved, as well as representatives of military industry, in attendance. Whilst the latter tried to maintain that everything should have worked as advertised and any failures were due to operator errors, in the end, certain improvements in combat systems, particularly the R-4 AAMs, were introduced as a result of the meeting.

Foreign aerostats aside, on at least one occasion, the Soviets were compelled to shoot their own ones down. In June 1974, unexpected air currents carried six Soviet balloons in the direction of China. In those times, the relations between the USSR and the PRC were strained and such a 'visit' by Soviet balloons could have been viewed by the Chinese as a provocation, even though none was intended. Thus, in order to prevent a possible international incident, the Soviets decided to shoot their own aerostats down. A Tu-128 of the 365th Fighter Aviation Regiment with Lieutenant Colonel Gaidukov at the controls, was scrambled and shot one down. Subsequently, aircraft flown by less experienced crews took to the air but failed to destroy the remaining balloons. This prompted the 365th Regiment's commander Colonel E. I. Kostenko, to personally deal with the problem at hand. He shot down another aerostat, demonstrating to his younger subordinates how it was to be done – whereupon, reportedly, they managed to bring the remaining balloons down.

Last Blast

Among the PVO's interceptors, the Su-15 proved to be a real balloon buster. On 17 October 1973, a trio of 62nd Fighter Aviation Regiment's Su-15s were scrambled to intercept an aerostat drifting at an altitude of 13,000m. The balloon's gondola was shot-off by means of a R-98 AAM, this becoming the first instance of a Su-15 firing in anger.[9] In the years which followed, Su-15s either shot down or incapacitated numerous other aerostats.

A particularly noteworthy incident of this kind took place on 3 September 1990. However, before relating those events, it is fitting to devote a few words to a seemingly different subject, namely the use of balloons for the purpose of scientific observation. During the 1980s, a new astronomical observation system was developed known under the acronym PIROG which stood for Pointed Infra-Red Observation Gondola.[10] The system compromised a high-altitude hydrogen filled balloon with a gondola containing observation equipment attached underneath. Due to the ability to reach high altitudes, astronomical observations of the interstellar medium in the infra-red spectral range could be performed above

Scramble! A photo showing a scene that happened several hundreds of times, every single year during the Cold War: pilot and his ground crew running to a Su-15 standing the QRA on one of air bases on the Kola Peninsula. (Albert Grandolini Collection)

A Su-15TM of the Besovets-based 57th Guards Fighter Aviation Regiment rolling for take-off while fully armed with R-98 and R-60 air-to-air missiles, in the mid-1980s. (via Easternorbat.com)

On the fringes of the USSR, there were three areas requiring particular attention of the PVO: Kola Peninsula in north-west, Caucasus area in the south and the Far East. Responsible for the defence of the Murmansk area and the northern Kola Peninsula, was the 21st Air Defence Corps PVO, headquartered in Severomorsk (also the primary naval shipyard, between others, responsible for constructing most of Soviet Navy's nuclear-powered submarines). The 21st Corps included a classic 'cell' of the Soviet Air Defence Force: three regiments of manned interceptors and five anti-aircraft missile regiments. One of its most important elements was the Kilpyavr-based 941st Fighter Aviation Regiment. Starting in 1959–1960, this was equipped with three squadrons of Su-9 interceptors – one of which is illustrated here as around 1968. Eventually, the unit replaced it with MiG-23Ms in the early 1970s. (Artwork by Tom Cooper)

Another of the units assigned to the 21st Air Defence Corps and responsible for air defence of the northernmost sector of the Kola Peninsula, was the Monchegorsk-based 174th Guards Fighter Aviation Regiment. For most of the 1960s, this comprised three squadrons, of which two initially flew MiG-19s (including the one responsible for the downing of an USAF ERB-47, on 1 June 1960), and another with Yak-25Ms. Starting in 1965, one of MiG-19-squadrons was re-equipped with Yak-28P interceptors armed with R-98 missiles (ASCC/NATO-codename 'AA-3 Anab'), one of which is illustrated here in livery around 1966–1967. (Artwork by Tom Cooper)

Deployed further east from the 21st Air Defence Corps and responsible for air defence of the Novaya Zemlya archipelago and the Vorkuta region, was the 4th Air Defence Division (HQ Belushya Guba). This comprised two regiments of manned interceptors and one anti-aircraft missile regiment: home-based at Amderma AB, was the 72nd Guards Fighter Aviation Regiment, which initially comprised three squadrons equipped with Yak-25Ms and MiG-19s. In the mid-1960s, the first of its units was re-equipped with Tupolev Tu-128s (ASCC/NATO-codename 'Fiddler') – at the time the biggest and most powerful interceptors of the PVO. Standard armament consisted of four Bisnovat/Molniya R-4R/T ('AA-5 Ash') air-to-air missiles: by 1975, all three squadrons flew that type. Another Tu-128-equipped unit deployed in the same area, was the Talagi-based 518th Fighter Aviation Regiment of the 23rd Air Defence Division (HQ Arkhangelsk). (Artwork by Luca Canossa)

On 20 May 1966, the PVO established the 67th Independent Air Squadron at the Monchegorsk Air Base. Initially equipped with just two Tupolev Tu-126 early warning and control aircraft, the unit was bolstered to six aircraft and then re-located to the Shavli AB, in the Baltic region later the same year. Although very limited in the look-down mode (where the range was less than 21km), its Liana radar complex had a detection range of 100km for aircraft the size of the MiG-17 and 300km for strategic bombers underway at the same or high-altitude. It closely cooperated with interceptor units of the PVO – especially Tu-128s – and greatly bolstered defence capabilities of the 10th Independence Air Defence Army. (Artwork by Tom Cooper)

The appearance of the Lockheed SR-71 Blackbird long-range, high-altitude, Mach-3-capable strategic reconnaissance aircraft of the 9th Strategic Reconnaissance Wing, USAF, over the Barents Sea, prompted the PVO to re-equip the Letnoezersky-based 524th Fighter Aviation Regiment of the 23rd Air Defence Division with two squadrons of MiG-25P interceptors, one of which is illustrated here. Its top speed of Mach 2.8 and armament of four R-40R/T air-to-air missiles (ASCC/NATO-codename 'AA-6 Acrid'; shown installed on the outboard pylon is an R-40R), proved insufficient of countering the Blackbird. Due to the defection of Viktor Belenko to Japan in 1976, in which all the secrets of the MiG-25P were revealed to the USA, the regiment had to be re-equipped with improved MiG-25PDS ('Foxbat-E') by 1980. (Artwork by Tom Cooper)

Based at the Afrikanda AB, the 431st Fighter Aviation Regiment flew MiG-15s from 1951 until 1955, MiG-17s from 1955 until 1960, and MiG-19P/PMs from 1960 until 1973, when it was re-equipped with Su-15. When a Boeing 707 of the Korean Airlines violated the Soviet airspace, around 20:54hrs local time on 20 April 1978, the 10th Air Defence Army ordered a scramble: with the nearest unit – the 941st Fighter Aviation Regiment – being in the process of re-equipping with MiG-23Ms, the task fell on the 431st instead. This is how Captain A. I. Bosov became involved in this incident. Notably, in the darkness, he misidentified the Boeing 707 as a Boeing 747, but still warned his superiors that it was a civilian airliner: nevertheless, he was ordered to open fire. The 431st Fighter Aviation Regiment continued flying Su-15TMs for the rest of the 1980s, and as of November 1990, still had 39 in operations. (Artwork by Tom Cooper)

During the mid-1980s, the PVO began introducing the practice of applying camouflage colours on some of its aircraft. The mass of patterns in question was applied 'in the field', included whatever colours were available and suitable for the purpose and – despite some attempts at regulations – applied in entirely different fashions, from aircraft to aircraft. This is the reconstruction of the Su-15TM flown by Captain Zhigulyov of the 54th GIAP during the encounter with Swedish SH.37 Viggens on 7 July 1985: the aircraft was painted in tan, dark brown, light green, and black or black-green on upper surfaces and sides, and wore a yellow Bort Number 36, outlined in white. The jet was armed with a pair of R-60s and R-98s, each. (Artwork by Tom Cooper)

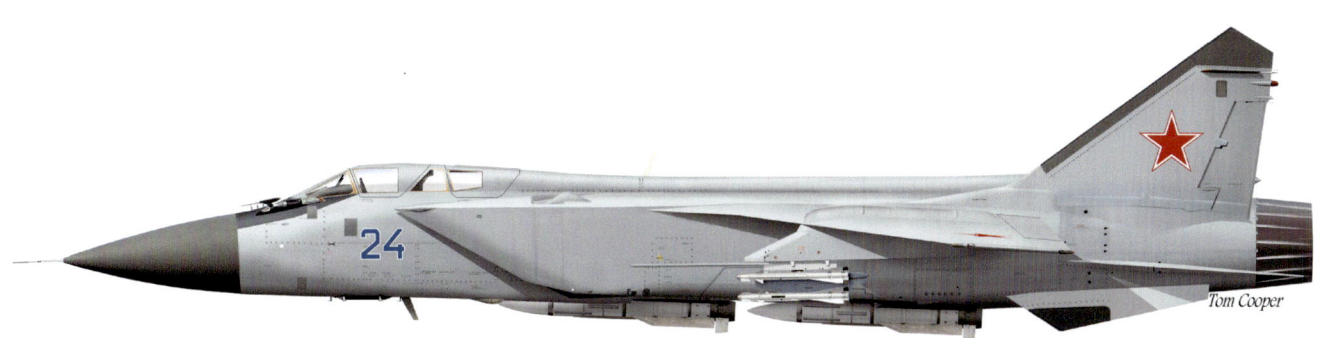

Starting in 1983, selected elements of the 10th Independent Air Defence Army began replacing their older types with the most powerful interceptor of the PVO ever: MiG-31. One of first units to convert to the new type was the Amderma-based 72nd Guards Fighter Aviation Regiment. Because operations from this isolated site proved excessively expensive, in October 1993, the unit moved to the Kotlas AB, where it was merged with the 445th Fighter Aviation Regiment. Nevertheless, the 72nd Regiment continued forward deploying its MiG-31Bs to Amderma whenever tensions with the West increased. This early MiG-31B of the 72nd was photographed by the Royal Norwegian Air Force in 1986, while armed with a full complement of four R-33 and four R-60 air-to-air missiles. (Artwork by Tom Cooper)

Although three of its regiments were still flying Su-15TMs, by 1987, the 10th Independent Air Defence Army included units flying MiG-31Bs and Su-27s, and ground-based air defence assets operating S-300PT/PS/PM (ASCC/NATO-codename 'SA-10A/D/E Grumble') SAMs. On 13 September 1987, Lieutenant Vasilly Tsymbal flew this Su-27P from the 941st Fighter Aviation Regiment, when intercepting the Lockheed P-3B Orion serial number 602 of the Royal Norwegian Air Force which was shadowing a group of Soviet Navy warships and a Beriev A-50 AWACS over the Barents Sea. Eventually, Tsymbal became involved in a light collision, from which everybody involved came out with a big scare. To avoid attracting attention, this Su-27P subsequently received the bort '38 Red', but also a 'kill' marking superimposing a P-3 in blue atop of a white star. Notably, like most of Su-27Ps of the 941st Regiment of this period, Tsymbal's Su-27 was armed with R-27ER/ETs only. (Artwork by Tom Cooper)

Starting in 1984, the PVO pressed into service the first out of eventually twenty-five Beriev A-50 early warning aircraft. Initially operated by the 67th Independent Aviation Squadron for Long-Range Radar Detection based at Šiauliai AB (re-organised as the 144th Aviation Regiment for Long-Range Radar Detection, in 1988), in the Soviet Republic of Latvia, the type's performance was only slightly better than that of the Tu-126: its Liana radar could simultaneously track 50–60 targets out to a range of 230km and control the work of about a dozen of own interceptors, but had a better look-down capability over the sea. Foremost, unknown to the Soviets was the fact that – just like in the case of the MiG-31 and Su-27 – all the secrets of the A-50 were compromised to the CIA by Adolf Tolkachev, one of top engineers of the Scientific Research Institute of Radar (NIIR or NII Radar, later Phazotron), in Moscow and before these types ever entered operational service. (Artwork by Tom Cooper)

The second area of focus for the PVO were the Black Sea and Caucasus, where British, US, Turkish, and Iranian reconnaissance and other aircraft frequently violated the Soviet airspace time and again during the 1960s and 1970s. Responsible for air defence of this area was the Baku Red Banner Air Defence District, most of which was still equipped with MiG-17s and MiG-19s as of the first half of the 1960s. Home-based at Salyan AB, in southern Soviet Republic Azerbaijan, and thus the closest to the border with the Iranian Empire, was the 627th Guards Fighter Aviation Regiment: as of 1964 this comprised three squadrons equipped with MiG-19S' and MiG-19P' interceptors (illustrated here). (Artwork by Tom Cooper)

During the 1970s, the Baku Red Banner Air Defence District was re-organised as the 19th Independent Air Defence Army – which had its HQ in Tbilisi, in Soviet Georgia. Nevertheless, the oil and gas producing Baku area remained heavily protected. Primary air defence asset in the area was the 82nd Fighter Aviation Regiment, home-based at Nasosnaya AB, outside the city. Formerly equipped with MiG-17s, since the mid-1960s, the 82nd flew Yak-28Ps as long-range, day/night, and all-weather interceptors. The unit was one of earliest to receive this 'late' sub-variant, with a longer radar nose and compatibility with R-3S ('AA-2 Atoll') air-to-air missiles in addition to R-8s. The 82nd replaced its Yak-28Ps by MiG-25Ps in the mid-1970s; by then its Yakovlevs had scored a handful of kills against so-called spy blimps. (Artwork by Tom Cooper)

The 982nd Fighter Aviation Regiment was not subordinated to the PVO, but units of the Tactical Aviation of the Soviet Air Force shared the burden of standing the QRA during the Cold War. This is how it comes about that on 28 November 1973, a MiG-21SM of that squadron – the reconstruction of which is illustrated here (obviously, the jet was left in bare metal overall) – intercepted an 'Iranian' RF-4C. According to Russian sources, the jet piloted by Captain Gennady Eliseev was armed with only two R-3S air-to-air missiles: considering the short-range of that type, it is possible that outboard underwing hardpoints were originally occupied by 490-litre drop tanks. In such a configuration, there was only enough space on the jet to install the two R-3S on its inboard underwing hardpoints. On order from the ground control, Eliseev then steered his MiG-21SM into a *taran* with the RF-4C, paying the ultimate price for – finally – downing one of the sneaky US-Iranian reconnaissance fighters. (Artwork by Tom Cooper)

Based at Sandar AB, outside Marneuli in the Soviet Republic of Georgia, the 166th Guards Fighter Aviation Regiment was re-equipped with Su-15s in 1970–1972. One of jets in question is shown on the main artwork as armed with R-98 ('AA-3 Anab') air-to-air missiles: the infra-red homing R-98T was always installed under the left wing, the semi-active radar homing R-98R under the right wing. After several incidents with low and slow-flying intruders, one in every pair of jets standing QRA was equipped with two UPK-23-250 gun pods with twin-barrel 23mm cannons under the fuselage. Initially short on qualified pilots and partially staffed by inexperienced novices from the Volunteer Society for Cooperation with the Army, Aviation, and Navy (DOSAAF), in the 1980s, the 166th partially re-equipped with Su-15TMs (insert). (Artwork by Tom Cooper)

Another Sukhoi-equipped unit of the Baku Red Banner Air Defence District was the 364th Guards Fighter Aviation Regiment, based at Nebit Dag, on the eastern coast of the Caspian Sea, which received its first Su-15s (inset) in 1971. Situated in the middle of a desert, with long, dry, hot summers, Nebit Dag was somewhat of a 'punishment colony' of the Soviet armed forces (and the juridical system), and nobody was looking forward to serving there. In 1976, the 364th was re-equipped with MiG-23Ms (main artwork, shown as armed with R-23T and R-13M air-to-air missiles), the first fighter jet with the lock-down/shoot-down capability in the Baku Red Banner Air Defence District. Three years later the regiment was merged with the 179th Guards Fighter Aviation Regiment, which continued operating MiG-23Ms into the late 1980s. (Artwork by Tom Cooper)

By the mid-1970s, Yak-28Ps were suffering alarming attrition rates and in urgent need of replacement. The 82nd Fighter Aviation Regiment was thus re-equipped with R-40-armed MiG-25Ps (main artwork): interestingly, its Yak-28P-crews were not converted to MiGs, but the unit received an entirely new group of pilots. Still based at Nasosnaya AB, the three squadrons of the 82nd Regiment thus became the southernmost air defence units operating this type in the entire USSR, and the MiG-25P '32 Red' (main artwork) was the regiment's top 'spy-blimp-killer'. In 1981–1982, personnel of the 210th Industrial Aircraft Repair Works overhauled and upgraded all of 82nd's MiG-25Ps to the MiG-25PDS-standard through the installation of the new N-005 Saphir-25 (RP-25M) pulse-Doppler radar with look-down/shoot-down capability. Jets like '21 Red' (insert) thus received the compatibility with short-range R-60 air-to-air missiles. (Artwork by Tom Cooper)

In the Far East, the PVO had the 11th Independent Air Defence Army (HQ Khabarovsk), comprising four major air defence groups. One of these was the 23rd Air Defence Corps (HQ Vladivostok), which during the 1970s and 1980s, included four regiments equipped with manned interceptors: the 22nd Guards Fighter Aviation Regiment (Tsentralnaya Uglovaya), 47th Fighter Aviation Regiment (Zolotaya Dolina), 530th Fighter Aviation Regiment (Chuguevka), and the 821st Fighter Aviation Regiment. The latter used to fly Bell P-39 Aircobra and P-63 Kingcobras during the 1940s and 1950s and MiG-17s until 1968. It was one of few PVO units to operate Su-7s, in 1965–1966. In 1968, it was re-equipped with Yak-28Ps, one of which is illustrated here, as armed with R-8T on its usual port underwing station. The regiment flew this type for 13 years, until re-equipped with MiG-23ML/MLDs, which it flew until being disbanded in 1994. (Artwork by Tom Cooper)

Another major element of the 23rd Air Defence Corps was the 530th Fighter Aviation Regiment: Established in 1951, three squadrons of this unit flew MiG-15s and MiG-17s until 1974, when all were re-equipped with MiG-25Ps. One of almost brand-new jets – Bort '31 Red' – was then flown by defector Senior Lieutenant Belenko to Hakodate, in Japan, in September 1976, delivering a major intelligence coup to the USA. This prompted the General Staff of the Soviet Armed Forces into launching a major upgrade of all the surviving MiG-25Ps to the MiG-25PDS-standard and placing orders for newly-built MiG-25PD. The 530th was thoroughly purged after Belenko's defection, but continued flying the type until 1990, when it was re-equipped with MiG-31s (main artwork). (Artwork by Tom Cooper)

On 1 May 1980, the PVO officially established the 40th Fighter Aviation Division with headquarters on Dolinsk-Sokol Air Base. This controlled the 308th Fighter Aviation Regiment equipped with MiG-21SMs (Burevestnik AB, on Iturup island), 528th Fighter Aviation Regiment with Yak-28Ps (Smirnykh AB), and the 777th Fighter Aviation Regiment with Su-15s (Dolinsk-Sokol AB). Piloted by Lieutenant-Colonel Gennady Osipovich, this Su-15TM '17 Red' of the 777th Fighter Aviation Regiment played the crucial role in shooting-down a Boeing 747 of the Korean Airlines on 1 September 1983. The 40th Fighter Aviation Division was disbanded on 1 May 1986 but meanwhile, re-equipped with MiG-31s, the 777th continued serving as an element of the 24th Air Defence Division (HQ Yuzhno-Sakhalinsk). (Artwork by Tom Cooper)

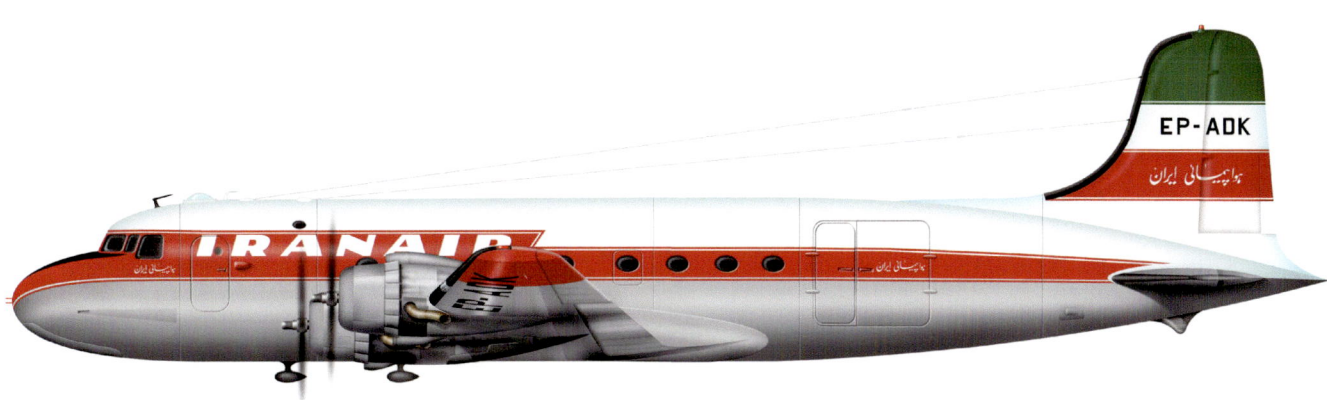

Ironically, after 1960, a growing number of PVO's targets – and victims – became innocent airliners or privately-owned light aircraft. One of the first was this Douglas DC-4 of the Iranian Airways, underway on a cargo flight from Beirut to Tehran, on 4 August 1961. The aircraft violated Soviet airspace due to a navigational error and was then attacked by a Soviet interceptor – probably a MiG-17 of the 82nd Fighter Aviation Regiment. The gunfire from the latter knocked out one and set on fire another engine but the crew managed a wheels-up landing inside Iran, near the southwest coast of the Caspian Sea. Because nobody was killed, and a delegation of top Soviet officials was in the process of visiting Tehran, the entire affair was quickly covered-up by both sides. (Artwork by Leon Manoucherians)

Exact details about the Phantom II rammed by Captain Eliseev on 28 November 1973, remain unclear: certain is only that the jet was one of a total of six RF-4Cs from USAF stocks, on loan to Iran – and not one of RF-4Es manufactured for Iran (eventually, the IIAF received a total of 12 RF-4Es). What is further known from eyewitness accounts, is that the jet, crewed by Major Shokouhnia and Colonel Saunders, wore the so-called 'South-East Asia' camouflage pattern in tan (FS30219), light green (FS34102) and brown-green (FS34079) on upper surfaces and sides, and light grey (FS36622) on under surfaces – which was never applied on any of IIAF-owned Phantoms. However, it wore the full Iranian insignia, including roundels and the fin flash. Whether it carried one of typical ECM-pods of the time – like ALQ-101, illustrated here – remains unclear, but IIAF RF-4s are known to have never ventured anywhere without one on-board. (Artwork by Tom Cooper)

Relations between the USSR and Turkey were always tense but especially once the latter joined NATO in 1952 and began providing multiple intelligence-gathering installations to its Western allies. Unsurprisingly, the PVO was always on alert whenever THK aircraft were passing close to its borders – which, often enough, was intentional: it prompted the Soviets to turn on all of their radars, thus exposing their dislocations and working frequencies. On 24 August 1976, this resulted in an incident in which PVO's S-125 SAMs ('SA-3 Goa') shot down an RF-5A reconnaissance jet of the 184. Filo. Sadly, the serial of the jet in question remains unknown but as of 1976, most of THK's RF-5As were painted as shown here, in tan (FS30219), light green (FS34102), and brown-green (FS34079) on upper surfaces and sides, and light grey (FS36622) on under surfaces. (Artwork by Tom Cooper)

Manufactured in 1966–1967, this Boeing 707-321B was operated by the Korean Air Lines with the registration HL7429. On 20 April 1978, while underway along a polar route from Paris via Anchorage to Seoul, it flew off course and penetrated the Soviet airspace down much of the Kola Peninsula. Two Su-15s were launched to investigate and force the jet to a landing in the USSR, but when the Boeing-crew appeared not to follow their instructions, the Soviets opened fire: one missile missed, the other severely damaged the port wing, forcing the crew to make an emergency landing on the Korpijärvi Lake. Whilst two passengers were killed by the damage caused by the missile hit, all other 107 occupants survived this incident. Eventually, the aircraft was abandoned in situ – especially once it became known that the Soviets ruined much of the fuselage while trying to find 'evidence' that it was on an 'espionage mission'. (Artwork by Leon Manoucherians)

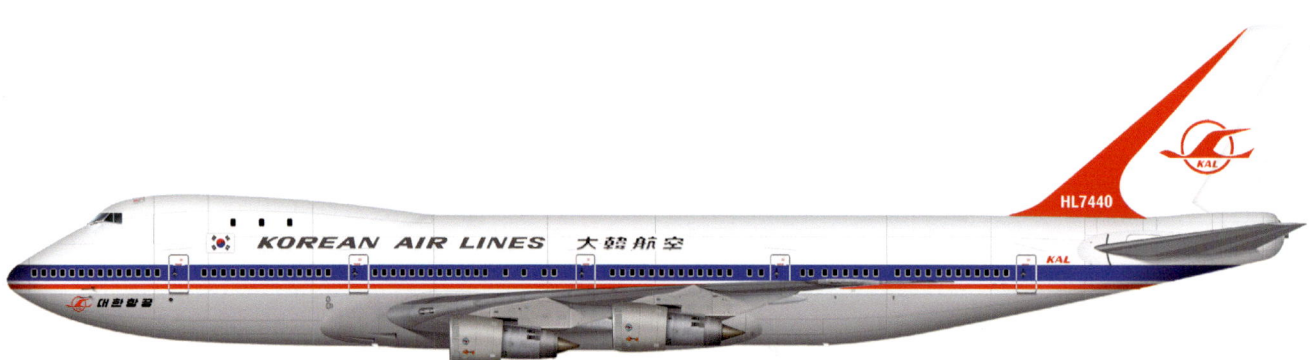

At the peak of the Cold War, on 1 September 1983, this Boeing 747-230B, manufactured in 1971 and operated by the Korean Airlines, flew into the Soviet airspace Pacific resulting in one of worst incidents in which military opened fire at civilians during peacetime. During that fateful night, this jet was underway on a scheduled passenger flight from New York to Seoul. After making a planned refuelling stop in Anchorage, it deviated from its planned route: as first, it passed over Kamchatka while pursued by at least five PVO interceptors; then it crossed the Sea of Okhotsk (considered 'Soviet' by Moscow) and entered the Soviet airspace again, this time off Sakhalin Island. At that point in time, it was intercepted by a Su-15TM of the 777th IAP and hit by a single R-98 missile. The latter caused sufficient damage for the jet to eventually crash into the sea near the Moneron Island, killing all 269 occupants (including 23 crewmembers). (Artwork by Leon Manoucherians)

A pilot of the 174th Guards Fighter Aviation Regiment posing next to his 'office' – in turn offering a good look at the instrument panels of the front and rear seat of his Yak-28P. Through the second half of the 1960s and into the 1970s, this was the most advanced, all-weather and night interceptor of the V-PVO. (Albert Grandolini Collection)

A Yak-28P basking in the sun at one of bases in the Baltic area of the early 1970s, while waiting for its maintenance crew. At the time, the type was operated by the 174th Guards Fighter Aviation Regiment (Monchegorsk, near Murmansk; 21st Air Defence Corps), 641st Guards Fighter Aviation Regiment (Besovets, near Petrozavodsk; 5th Air Defence Division), and the 524th Fighter Aviation Regiment (Letnoezerskiy/Obozersk; 23rd Air Defence Division). (Albert Grandolini Collection)

A front view at a Yak-28U. Whilst seeing no 'action', these two-seat conversion trainers played a crucial role not only with helping convert new pilots and ground crews to the type, but also through providing continuation and refresher training for experienced aircrews. (Albert Grandolini Collection)

A pilot of a Tu-128 interceptor posing in front of his mount. Notable are not only the 'business ends' of two R-4 air-to-air missiles to the left, but his partial protective suit and heavy boots, worn atop of the g-suit – badly necessary in the case of an emergency over the Siberian expanses. (Albert Grandolini Collection)

A Su-15TM of the 57th Guards Fighter Aviation Regiment approaching for landing at the Alykel AB, outside Norilsk, in the Polar Circle. Two squadrons of this regiment were permanently deployed at this facility starting in 1987: although obsolete since the early 1970s, Su-15 was available in relatively large numbers, reliable and possessed a good operational range – which is why it continued serving well into the 1980s. (Albert Grandolini Collection)

A Su-15 streaming the braking parachute on landing. Notable is the installation of the hardpoint for R-60 air-to-air missiles near the wing root: high reliability and availability of this type made sure that even this earlier variant continued soldiering into the 1980s, with only minimal modifications. (Albert Grandolini Collection)

A Su-15TM, fully armed with a pair of R-98RM and R-98TM air-to-air missiles, and a pair of short-range R-60M missiles, as seen during an encounter with a NATO aircraft in the Baltic of the early 1980s. (Albert Grandolini Collection)

Starting in 1985–1986, NATO air forces underway over the Barents and the Baltic Seas, began encountering the most powerful interceptor of the PVO ever – in form of the MiG-31B. This example was photographed not fully armed. It has the launch rails for the R-60 AAMs but not the missiles themselves. (US DoD)

One of several Su-27Ps of the 941st Regiment, encountered by P-3Bs of the Royal Norwegian Air Force in September 1987, before a light collision between two aircraft brought all such 'games' to an end. Notable is the heavy medium-range armament of this jet, in form of four R-27ER semi-active radar homing and two R-27ET infra-red homing air-to-air missiles. Launch rails for the then, brand-new, R-73 short-range missiles were in their place, but no missiles installed on them. (Royal Norwegian Air Force)

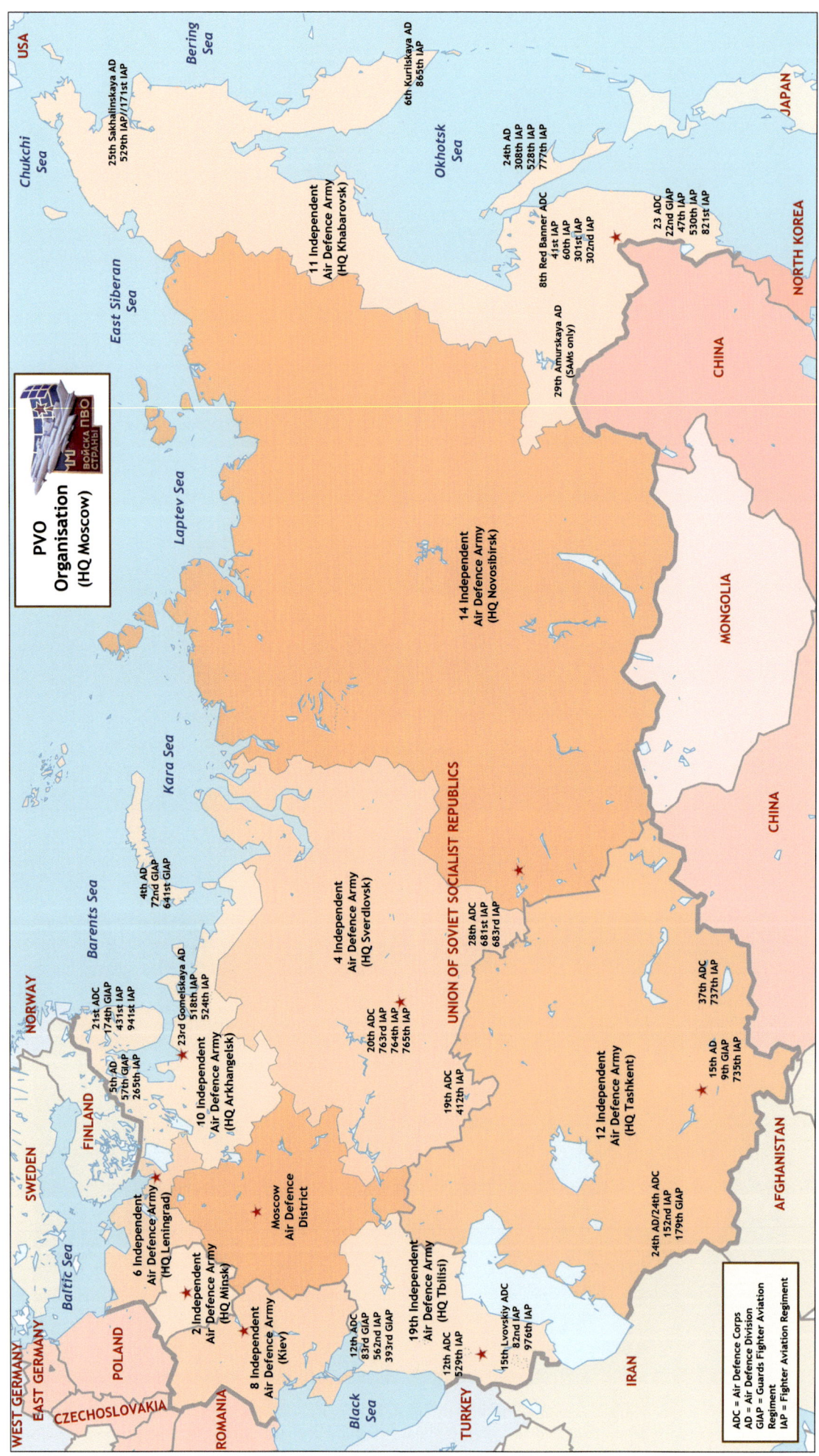

A map of the areas of responsibility of the PVO as of the 1980s. (Map by Tom Cooper)

Eventually, Su-11s of the PVO never engaged their planned opponents, nor any of so many violators of the Soviet airspace. However, one shot down a Soviet Su-9 during training, and another scored a kill against a so-called 'spy blimp'. (Albert Grandolini Collection)

the lower layers of the atmosphere by the PIROG. Once a mission was accomplished, the flight was terminated by cutting the balloon line via ground command with the gondola descending to earth by means of parachutes.

Four flights were successfully performed, and nothing indicated that the fifth one appropriately, named PIROG 5, which was launched on 27 August 1990 from the European Space Range site in Kiruna, Sweden would be any different. However, for reasons not known, the separation system by radio command failed. Thus, the out-of-control aerostat with the gondola attached, drifted over the Arctic Sea floating into Soviet airspace in the vicinity of Murmansk by 2 September 1990.

The balloon was tracked by Soviet AD radars and it was decided to deal with it as with numerous other foreign aerostats which had entered Soviet airspace in times past, that is to shoot it down. However, the decision was also made not to down it with surface-to-air missiles, even though this could have been done with ease. There were a number of reasons for this: first of all, after firing the SAMs and hitting the target, their falling debris could pose a hazard; in addition, it seemed advisable to visually inspect the balloon and its payload. Finally, it was prudent to ensure that the balloon was not manned, even if the possibility was remote.

Thus, the aerostat floated unhindered in the vicinity of Murmansk giving the inhabitants of the city and surroundings, many visual thrills for as the lighting conditions, changed so did its colour. Finally, the balloon drifted out of sight in the southern direction. Once it was over sparsely inhabited areas, its fate was sealed. A Su-15TM of the 431st Fighter Regiment took off from Afrikanda with Captain Igor Zdatchenko at the controls, tasked with intercepting and downing the aerostat.[11] The actual shoot-down took place some 50km north-west of the town Kovdor at an altitude of ca 14 000m, with the date and time being 3 September 1991 and 08:08 respectively.[12]

Two R 60 air-to-air missiles were launched by the Soviet fighter, one of which struck the balloon's envelope. The fact it had enough of a thermal signature for the IR guided missile to home on to, is in itself, noteworthy. The missile's warhead detonated, shredding the aerostat's envelope and sending its gondola plunging earthwards. However, soon the parachutes deployed and thus the gondola floated down, coming to rest intact on the ground.

The incident had an interesting aftermath. Not surprisingly, the Swedes requested the gondola to be returned but the Soviets' official reply was, that the item came down in a swamp and sunk into it. Meanwhile however, Soviet scientists had written to their Swedish colleagues about the coming down of the scientific balloon's gondola and enclosed a photograph of it with the letter.[13] Once 'armed' with ad oculus evidence of its survival, the Swedes took the matter to the Soviet authorities for a second time. Since the evidence was irrefutable, the Soviets could not deny the possession of the gondola and eventually returned it but according to the Swedes, removed a number of high-tech items first. It should also be added that the incident described did not derail the PIROG programme which subsequently continued with more balloon launches.

It seemed Captain Zdatchenko's shoot-down related above was the last PVO fighter aviation's Cold War 'kill' however, as it turned out that, honour goes to Captain Ivanov of the 62nd IAP. Piloting a Su-15 on 19 January 1991 at 10:41 in the area Sudak-Kirovskoye-Sovetsky (Crimea), he shot down an aerostat floating at an altitude of 18,500m (60,696ft).[14] Unless new data comes to light, this can be considered the PVO's Cold War last blast.

In addition to the ones described previously, there were other Soviet balloon shootdowns, for example, the Su-11 had at least one such 'kill' under its belt.[15] Last but not least, it is also worthy of note that Cold War balloon busting was bloodless, which was very fortunate since other aerial incidents tragically led to heavy loss of life as was the case with passenger aircraft.

7
THE SAFEST MODE OF TRANSPORTATION

Among many aircraft the PVO's fighter aviation intercepted or shot down, sadly there were also airliners. The first incident of this sort seemed to have taken place on 8 January 1962 when a Sabena Belgian Airlines Sud Caravelle was forced to land at Grozny in the Caucasus region of the Soviet Union by four MiG fighters.[1] The Caravelle was on a flight from the Iranian capital Tehran to Istanbul in Turkey when it encountered bad weather, suffered a compass failure and as a result, inadvertently penetrated relatively deep into Soviet airspace. Subsequently, the passengers as well as the aircraft were released.

There were also at least two cases of Soviet fighters intercepting 'red' passenger aircraft: a Soviet one on a domestic flight, as well as a Hungarian Tupolev coming from abroad.[2] Little is known about either of those two events, but it seems neither brought about really serious repercussions. Not every incident had such a fortunate ending. As already mentioned, the crash of a Tu-104 passenger aircraft on 30 June 1962 some 28km east of Krasnoyarsk airport, resulting in 82 fatalities (all who were aboard), may have been caused by a SAM fired in course of a PVO live firing exercise, but this was never officially confirmed.

Douglas Dunking

In the morning hours of 1 July 1968, an intrusion into Soviet airspace in the vicinity of the Kuril Islands was detected, prompting an emergency scramble by a pair of MiG-17s from the 308th IAP, with Captain Aleksandrov and Captain Igonin in the cockpits.[3] The airspace violator was intercepted and turned out to be an American Douglas passenger aircraft. Initially, signalling by Soviet pilots was ignored by the Americans. It was only after two more MiGs, flown by Major Yevtushenko and Captain Moroz, joined the first duo that the airliner's pilot became responsive. Eventually, the American aircraft was forced to land on Burevestnik airfield, located on Iturup Island. As it transpired, the aircraft in question was a Seaboard World Airlines Flight 253A, served by a Douglas DC-8 Super 63CF. Of the 238 people who were aboard, 214 were US servicemen for it was a military charter flight carrying American troops from the USA via Japan to South Vietnam.

On the afternoon of 1 July 1968, the DC-8 departed McChord Air Force Base, near Seattle, Washington bound for Yokota Air Base in Japan. Unfortunately, the aircraft strayed westward of its planned course, bringing it into the vicinity of Soviet Kuril Islands. Japanese air traffic controllers notified the Americans of the error as they came within the range of Japan's ATC, yet the message was either unintelligible or came too late. In any case, the DC-8 was intercepted and forced to land by PVO fighters. Conditions at Burevestnik were austere, even by Soviet standards, but the Americans would not be stuck there for long.

Fortunately, the Nuclear Non-Proliferation Treaty, which had been negotiated some weeks earlier, had just been signed by the US President Lyndon Johnson. This created a moderately positive atmosphere between the superpowers and under these circumstances, there was no desire on USSR's behalf to escalate the situation.[4] Just how lucky the Americans were can only be seen in historical context. Namely, 1968 was the year of the Tet Offensive in Vietnam, as well as the year when the North Koreans sized the USS Pueblo. Thus, the Cold War was raging, with a lengthy and involuntary stay in the Soviet Union being a realistic possibility for the Douglas' crew and passengers. Yet, due to the circumstances explained, the Soviets were content when the US issued a short note of apology and the DC-8's pilot Captain Joseph Tosolini also apologised for violating USSR's airspace, allowing the US aircraft with all the Americans to depart.[5]

Boeing Busting

Whilst the incident described ended without injury or loss of life, it was not always the case. As with many other similar events, everything started to unfold when PVO's radars detected an unidentified aircraft which was still far out but heading for Soviet territory.[6] The one appearing on the radar screens in the evening hours of 20 April 1978, would enter USSR's airspace somewhere in

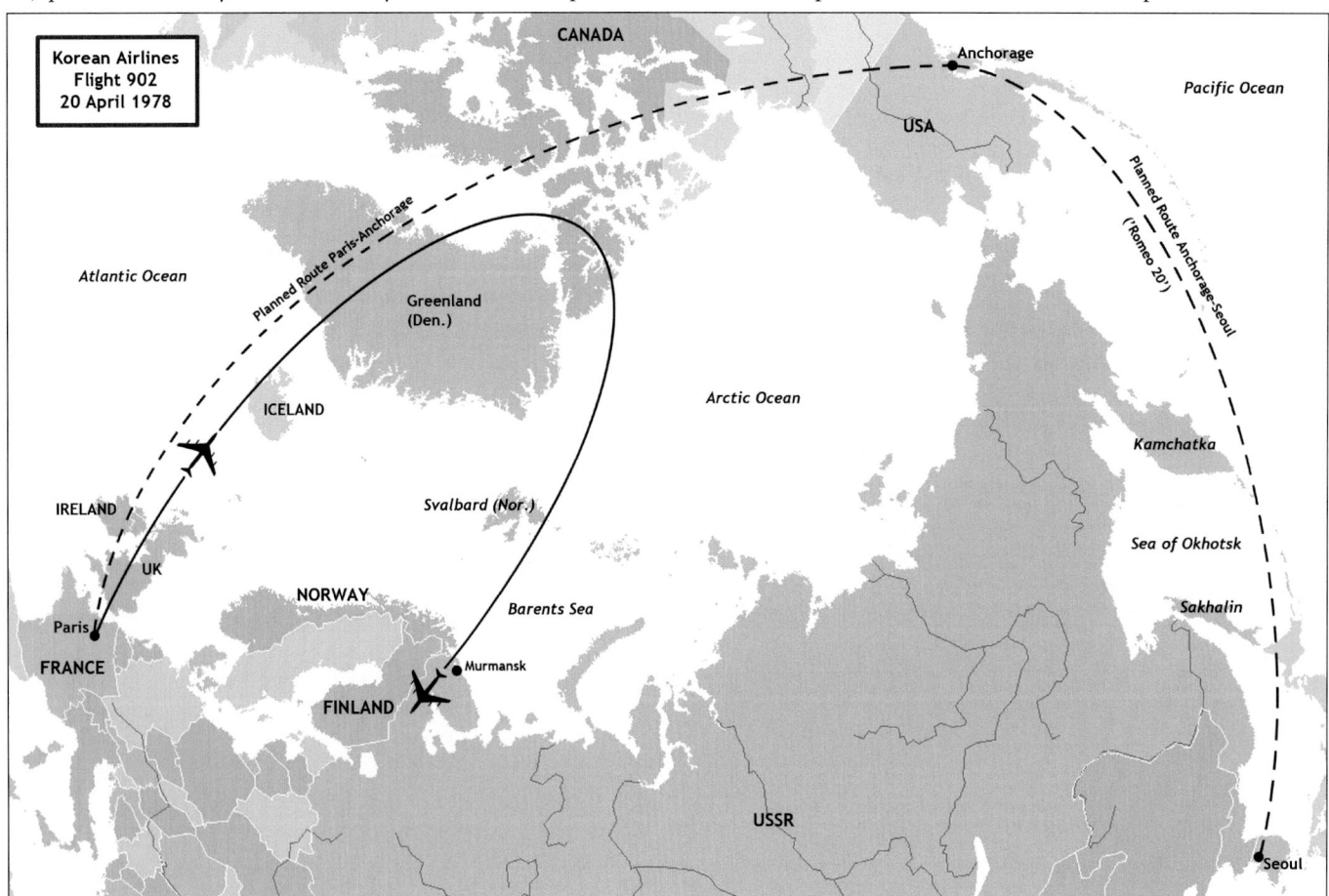

The route the Flight 902 of the Korean Airlines was supposed to take and the route it actually took which brought it over the Kola Peninsula, right in front of two pairs of PVO's Su-15TMs. (Map by Tom Cooper)

The Korean Airlines Boeing 707-321B registration HL7429, as seen on the morning after its shoot-down. (Albert Grandolini Collection)

A 15TM of the 431st IAP, as seen in the early 1980s. Notably, while the first pair of Sukhois lost the B707 as this descended into the clouds, another pair of Su-15TMs from the 265th IAP approached the scene: one of them even fired an R-8R missile at a 'slow-flying target' – which turned out to be the debris from the badly damaged airliner. (Albert Grandolini Collection)

the Severomorsk – Murmansk area if it kept its course. The situation warranted to scramble a fighter and accordingly, a Su-15 of the 431st Fighter Aviation Regiment with Captain Bosov at the controls, took off from Afrikanda. By the time the foreign aircraft was intercepted, it had already entered Soviet airspace.

Captain Bosov reported the aircraft to be a Boeing 747 with writing in Chinese, Japanese or Korean characters on its fuselage. As a matter-of-fact, the aircraft in question was a South Korean Boeing 707 as will be explained. Captain Bosov was ordered to force the airspace violator to land, whereupon he repeatedly rocked the Sukhoi's wings. However, the foreign aircraft's pilot was unresponsive to signalling. Moreover, the airspace violator was meanwhile, nearing the Finish border. As things stood and despite the fact that the intercepted aircraft was an airliner, the Soviet pilot was ordered to shoot it down.[7]

Reluctantly, Captain Bosov launched a R-98 AAM which struck the Boeing's left wing, blowing off its outer part. Unfortunately, fragments produced by the missile's explosion perforated the fuselage causing decompression in the cabin and mortally wounded two passengers: Bahng Tais Hwang and Yoshitako Sugano (a Korean and Japanese national respectively). The stricken airliner rapidly descended and was lost by Soviet radars. Luckily, its pilot sighted a large swath of empty space suitable for an emergency landing. As it turned out, it was the frozen surface of Lake Korpijärvi and there, the Boeing made a successful belly landing.

Meanwhile, the Soviets had scrambled more interceptors, among them another Su-15 of the 265 IAP, flown by Senior Lieutenant Slobodchikov, who made radar contact with one more target and launched a R-98 AAM at it. Subsequently it turned out that the 'target' was fragments of the Boeing's shot-off wing tip which while swirling in the air, produced a return on radar screens.

The aircraft at the centre of this incident was a Korean Air Lines Boeing 707-321B (reg. HL7429). On 20 April 1978, it departed Paris in France for the Flight 902 (KAL 902) to the South Korean capital Seoul, via Anchorage in Alaska with a total of 109 persons aboard. The Korean Boeing passed in the vicinity of the Canadian Forces Station 'Alert' located on Ellesmere Island, where the crew corrected their course.[8] However, this 'correction' was in fact, a turn in the wrong direction which brought them on a course across the Barents Sea, directly towards the Soviet airspace. From that point, the events unfolded as already related. After about two hours, Soviet military personnel arrived at the scene of the aircraft's emergency landing.

On the next day, 21 April, all those who were aboard were evacuated by helicopters to the town of Kem in Soviet Karelia. Except the pilot Kim Chang Kyu and navigator Lee Kun Shik, the rest of the crew, the passengers, as well as the bodies of the deceased, were released via Finland on 22 April. According to official Soviets sources, both the pilot, as well as the navigator, had confessed to violating Soviet airspace and disregarding orders from intercepting aircraft to land.[9] Both appealed to the Presidium of the Supreme

Soviet for clemency whereupon Soviet authorities decided not to criminally prosecute them but instead expelled the two men from the USSR on 29 April. Once out of the Soviet Union, they claimed to have reduced speed and switched on the landing lights when intercepted.[10] Thus, they supposedly showed their willingness to comply with whatever Soviet fighters would have ordered them to do yet, the latter opened fire none the less.

In an aftermath to the whole affair, the Soviet Union demanded US$100,000 from South Korea for costs resulting from the incident, which rather unsurprisingly, the South Koreans refused to pay. As for the Korean Boeing 707, it was left at the site of its landing. How the PVO command viewed the incident was both interesting and revealed plenty about it. Namely, bringing down an airliner whilst obviously, an unwelcomed event, was not the worst thing that could have happened, for from its perspective, it would have been much worse had the airspace violator managed to get away. As a result, no-one was rewarded for his actions that day, but nobody was punished either, since in essence, those involved did their duty. Sadly, this incident resulted with loss of life but a tragedy far worse by several orders of magnitude was yet to come.

Target is Destroyed
Korean Air Lines Flight 007 (KAL 007) was a scheduled passenger flight from New York, USA, to the South Korean capital Seoul via Anchorage, Alaska.[11] The aircraft, a Boeing 747-230B (reg. HL7442) with 269 people aboard (a 23 men crew and 246 passengers), arrived at Anchorage at 11:30 UTC (03:30 local time). A crew change took place and the airliner was readied for the final flight leg to Seoul. At 12:58 UTC KAL 007 received clearance for take-off from Anchorage and was airborne at 13:00. Thus far, the flight was uneventful and everything went according to a well-established routine. However, about 10 minutes after take-off, KAL 007 began to deviate to the north of its assigned route and continued to fly on this constant heading for the next five and a half hours. This resulted in KAL 007's progressively greater lateral displacement to the north of its planned flight route, which ultimately led the Korean Boeing to penetrate Soviet airspace first overflying Kamchatka Peninsula and subsequently Sakhalin Island as well as the surrounding territorial waters.

While KAL007 was flying through the darkness over the Pacific Ocean, PVO radar operators on Kamchatka were busy tracking Target 6064 which was apparently, an American RC-135 reconnaissance aircraft operating over international waters.[12] At 04:51hrs local time, another contact appeared on the radar screens which was assigned the number 6065. It was flying at an altitude of 8,000m with a speed of 800km/h. Initially, the radar operator thought it was an aerial tanker send to replenish the RC-135's fuel. However, the new contact flew past the RC-135 and headed straight for Kamchatka, entering Soviet airspace at 05:34. A Su-15 of the 865th IAP was scrambled from Elizovo but failed to intercept the airspace violator before it left the Soviet airspace over the Kamchatka Peninsula.

Meanwhile, alarm was sounded at Dolinsk-Sokol airfield where the 777th IAP was based. There, a Su-15TM '17 Red', armed with two R-98 AAMs, as well as two gun pods, was on QRA. It was manned by Lieutenant Colonel Genady Osipovich who was scrambled at 05:42, with the call-sign 805. In addition, a MiG-23 of the 528th IAP, flown by Major Litvin, was scrambled from Smirny AB and then another Su-15TM of the 777th IAP, piloted by Major Tarasov.

The ground control decided to vector all three fighters on a rear conversion course. The one to catch up with the target, which unbeknownst to the Soviets was the Korean Boeing 747 KAL 007, was Lieutenant Colonel Osipovich. At 06.03hrs, ground control informed Lieutenant Colonel Osipovich that the target was about 12-15km ahead of him. A little over 10 minutes later, at 06:14 the Soviet pilot was instructed to prepare to engage it. At 06:17, ground control first enquired if the pilot had acquired the target and having received an affirmative reply, ordered it to be destroyed. Lieutenant Colonel Ospovich requested the order to be repeated, whereupon at 06:18, he received the reply that the target had crossed the Soviet border and was to be destroyed.

Indeed, the target, that was the Korean Boeing, penetrated into Soviet airspace over Sakhalin and from that moment on, the events accelerated towards a dramatic climax. Meanwhile, ground control inquired if the intercepted aircraft's navigational lights are on, to which Lieutenant Colonel Osipovich replied that they were and that a Strobe light was flashing as well. At 06:19, ground control instructed the Soviet pilot to switch on his navigational lights and force the intruder to land on his home-base while the Soviet pilot informed ground control that his AAMs' seeker-heads had already acquired the target.

At 06:20, ground control ordered him to fire warning shots and in fact, the Soviet pilot fired hundreds of cannon shells but apparently, to no effect. One minute later, at 06:21 Lieutenant Colonel Osipovich informed ground control that the target had decreased speed and as a result, he found himself ahead of it. At 06:22 ground control ordered the Soviet pilot to fire at the target, to which he replied that this should have been done earlier but soon assumed a position behind the targeted aircraft. Two minutes later, at 06:24, ground control advised Lieutenant Colonel Osipovich to try to destroy the target with gunfire but the Soviet pilot stated that he will use missiles.

At 06:25, ground control ordered him to fire, to which Lieutenant-Colonel Osipovich replied that he launched the missiles, reporting one minute later at 06:26, that the target was destroyed.

Aboard the Boeing 747 flight KAL 007, no-one was aware of the situation.[13] Obviously, the crew did not know its actual position, otherwise the Koreans would not have flown where they did. In addition, they also did not notice Soviet interception attempts, including the firing of warning shots and the flashing of lights. This should not be surprising though, as the view from an airliner's cockpit is rather limited and the Sukhoi's guns were not loaded with tracer ammunition.

While a Soviet fighter-interceptor was about to shoot them down, KAL 007's crew was engaged in casual conversation on the flight deck and in an exchange via radio with the crew of KE015, which was another Korean Air Lines flight to Seoul. Both aircraft were also in contact with Japanese air traffic control. Shortly after 18:20hrs, UTC, KAL 007 commenced a planned climb manoeuvre, which for a short time, spoiled Osipovich's interception attempt (see earlier). At about 18:25 UTC the Boeing was hit, likely by just one of the Soviet AAMs: a hole with a total area of about 0.16 square metres was blown open in the fuselage, causing decompression; moreover, the aircraft's hydraulics and with them the control systems, were seriously damaged. Due to the damage sustained, the airliner pitched up and initially accelerated vertically also rolling slightly, right wing down. Eleven seconds after the hit, the sound of the cabin altitude warning was heard in the cockpit. As a result of hydraulic systems' failures, due the aforementioned damage, the aircraft became increasingly difficult to control. The aircraft continued to climb and reached a maximum altitude of 38,250ft (11,658m) losing speed in the process. Subsequently the aircraft started to descend, temporarily recovered to level flight, possibly stalled and then

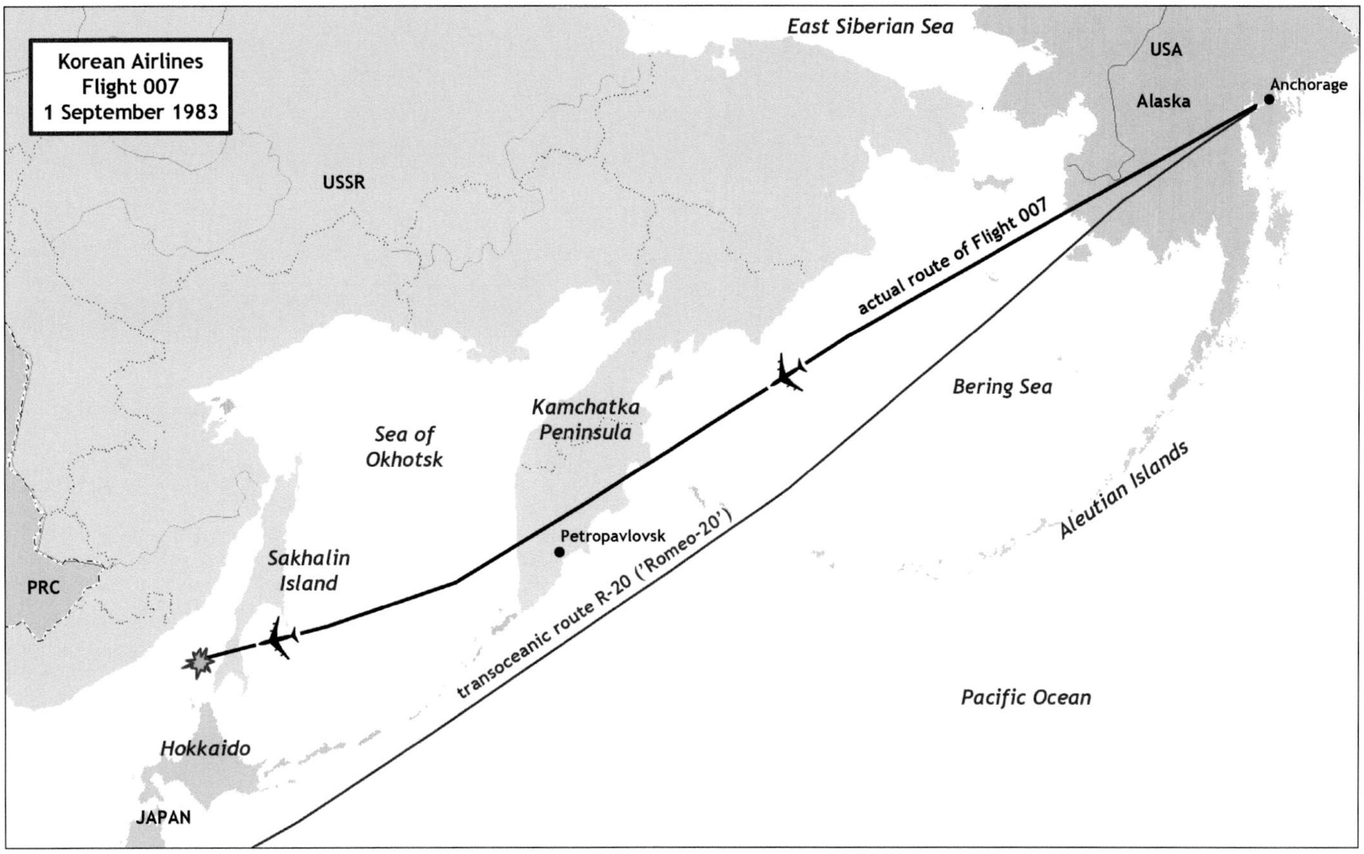

A map of the tragic Flight 007. (Map by Tom Cooper)

spiralled down to the left until it impacted on the surface of the sea. As a result of the crash, all those aboard, that is a total of 269 persons, lost their lives.

This – rather dry – account merits a few additional words. Once the Soviet Union was no more, Gennadiy Osipovich, who had retired with the rank of Colonel (he passed away, meanwhile), spoke to the domestic, as well as the foreign press.[14] Doing so, he revealed a number of interesting details which add to the understanding of this particular event and arguably, also give a more general insight into the way a PVO fighter-interceptor pilot thought and acted. While it was speculated that the Soviet pilot might have mistaken a civilian airliner for a military reconnaissance aircraft, it was not the case. Osipovich told interviewers that he saw two rows of windows, knew that this was a Boeing and also knew that it was a civilian aircraft. Yet for him this meant nothing for he rationalised that it was easy to turn a civilian type of aircraft into one for military use.

Moreover, he did not provide a full description of the aircraft to ground controllers. Specifically, Osipovich did not tell them that it was a Boeing-type aircraft and again, he rationalised this by stating that he was not asked about it. The Soviet pilot did not try to establish radio contact with the intercepted aircraft, once more explaining away his actions: there was no time for that and besides there was the language barrier. However, Osipovich did use his fighter's IFF

On 1 September 1983, it was once again that Su-15TMs of the PVO become involved in an intercept of a Korean Airlines passenger aircraft that flew off the route. Contrary to his colleagues over the Kola Peninsula, five years earlier, Lieutenant Colonel Osipovich did not care to inform his ground control that the jet he had intercepted was a civilian airliner: instead, he followed his orders and opened fire. Whilst no photographs of his jet became available, this Su-15TM was photographed while underway in exactly the same weapons configuration: R-98T under the left wing, two UPK-23-250 gun pods under the fuselage and an R-98R under the right wing. (Albert Grandolini Collection)

system to enquire if the intercepted aircraft was not a friendly one. Since obviously a Korean airliner was not equipped with a Soviet IFF transponder, no reply was received and the aircraft was thus, a hostile one with all consequences this entitled. The Soviet pilot did not harbour any murderous intent and would have preferred to have forced the foreign aircraft to land. Yet, viewed through the prism of the Cold War, Osipovich treated the aircraft, not as an airliner which lost its way, but as part of a nefarious undertaking against his motherland with the latter justifying its physical destruction. Thus, Genady Osipovich had no regrets and if anything, he was angered by the fact that for all his trouble, he was paid only a modest bonus of 200 Rubles, minus a small postage fee.

The matter did not end with Korean Airlines Boeing 747 flight KAL 007's destruction but, in a way at that point, the whole affair just started. American monitoring stations at Elmendorf, Alaska in the

Ironically, both intercepts of Korean Airlines aircraft in 1978 and 1983, fully exposed a massive weakness of the Su-15TM's Taifun-M radar and fire-control system: this could not detect any kind of targets flying lower than the Sukhoi. Unsurprisingly, all of the involved pilots experienced massive problems with finding their targets or tracking them over extended periods of time – and lost them in the darkness as soon as the two Boeings descended to lower altitude. This busy scene was taken around the same time, during an exercise. (Albert Grandolini Collection)

United States, Misawa in Japan and Wakkanai also in Japan, recorded PVO's communications immediately prior and during the shoot-down.[15] The tapes with the said recordings were promptly collected and by 2 September, were physically delivered to continental United States where their analysis was swiftly undertaken. Soon enough, the Americans put together a transcript of PVO's communications which gave them a picture of what happened to Korean Airlines Boeing 747 flight KAL 007. They at once realised the importance, as well as value of the materials in their hands and decided to make use of them.

On 3 September, US President Ronald Reagan was briefed about the matter and on 5 September, he gave a televised speech condemning the shoot-down, calling it the Korean airline massacre, a crime against humanity that must never be forgotten, an act of barbarism and inhuman brutality.[16] The Soviets initially issued a denial and subsequently, made announcements which contained a number of inaccuracies, as well as outright falsehoods. Namely, the Soviets claimed that an aircraft which deeply penetrated into USSR's airspace, was flying without navigational lights, that it was intercepted by PVO fighters but failed to follow their instructions and finally, that it left Soviet airspace, after which track of it was lost.[17]

A few days later, the Soviets claimed that the foreign aircraft acted in coordination with an American RC-135 reconnaissance aircraft, that among others, efforts were made to contact it via radio and that in the end, a PVO fighter-interceptor was ordered to terminate its flight in accordance with relevant Soviet laws.[18] During a press conference held on 9 September, Marshal of the Soviet Union Nikolai Ogarkov, reiterated the accusation that KAL 007's flight was a deliberate provocation, going as far as to assert that the heavy loss of life resulting from its destruction may have been something those who concocted it, had hoped for.[19]

However, Soviet efforts to deflect blame for the tragedy were by large, unsuccessful and in the court of the world's public opinion, they were almost universally condemned as the guilty party. In contrast, the Americans played the whole thing as close to perfection as possible, scoring a major propaganda victory against their Cold War foe.

Yet another aspect of KAL 007's destruction, was that it produced physical remains in the form of flotsam as well as wreckage. While the former could be picked up from the surface of the sea, the latter sunk to its bottom. Locating and retrieving the wreckage was of considerable importance as the flight recorders would be among it. Thus, a race between the Soviets and the Americans to find the wreckage started with Soviets having a lead – since they had shot the aircraft down, they had a better idea were to start searching for its remains. This part of KAL 007's story is interesting, in itself meriting a lengthy description but sufficient to say, the Soviets were indeed the first to locate the Korean Boeing's wreckage and to retrieve its flight recorders. However, they were uncooperative with the International Civil Aviation Organization (ICAO) investigation of this matter and did not share their findings.

It took events of historic proportions as the USSR attempted to reform, only to dissolve in December 1991, for things to change in this regard. In autumn 1992, the Russian President Boris Yeltsin, ordered the declassification of a number of Soviet documents pertaining to this matter which were published by the Russian press.[20] Subsequently, more documents, as well as other items, were released by the Russians with the tapes of the cockpit voice recorder and the digital flight data recorder of the Boeing 747 KAL 007 handed over to the ICAO in January 1993.[21]

Last but not least, the question of how the Korean Boeing could find itself in Soviet airspace deserves to be answered. As already stated after take-off, KAL 007 began to deviate to the north of its assigned route and this resulted in a progressively greater lateral displacement from its planned flight path which ultimately led it to penetrate Soviet airspace. Obviously, it begs the question, how did this deviation come about? Analysis of the flight data by the ICAO determined that the said deviation was probably caused by the aircraft's autopilot system operating in the HEADING mode, after the point that it should have been switched to Inertial Navigation System (INS) mode.[22] The HEADING mode maintained a constant magnetic course selected by the pilot. In contrast, when the INS navigation system was properly programmed with the filed flight plan waypoints, the autopilot's mode selector switch could be turned to the INS position and the aircraft would then automatically track the programmed INS course line, provided it was headed in the proper direction and within 7.5 miles (12.1km) of that course line.

Whatever the reason, the autopilot remained in the HEADING mode and the problem was not detected by the crew. Thus, due to the aforementioned deviation after take-off and the autopilot working in the HEADING mode, the aircraft flew straight ahead with the displacement from its planned course steadily growing, which brought about the consequences.

Closing the matter, it needs to be pointed out that only the most important facts were related herein. Those interested to find out more can refer to numerous films, TV reportages, books as well as articles which were produced on this subject. However, the potential viewer and reader needs to be aware that many such contain various inaccuracies or make assertions, for which solid evidence is lacking. Fortunately, a host of source material has also been declassified since and is nowadays available for anyone seeking factual information regarding this tragic, as well as controversial, Cold War incident.

Fortunately, after KAL 007's demise there were no more incidents involving the PVO and civilian airliners which resulted with loss of life, however, a few, luckily bloodless ones, still took place. According to the Swedish government on 9 August 1984,[23] a Soviet fighter pursuing an Airbus 310 jetliner, intruded 30 miles into Swedish airspace, at one point closing to within about a mile of the airliner, which was unaware of the fighter. Radio intercepts showed that the Su-15 fighter-interceptor had locked-on the airliner its air-to-air missiles. It appears Su-15s belonging to the 54 GIAP were engaged in an exercise conducting mock interceptions of Tu-16 (ASCC/NATO-codename 'Badger') bombers acting as 'targets'. Likely, GCI operators confused the Airbus with a Tupolev and this is how the incident came about. The Swedish government treated the incident very seriously since its airspace was violated, but due to the nature of the event (an exercise), it seems the risk of the passenger aircraft being actually fired at, was very low. In any case, on 21 October, the Soviets officially denied that any such thing had happened and claimed the jet was 50 miles from where the Swedish radars showed it.

Apparently, the last incident involving passenger aircraft and PVO fighters during the Cold War took place on 29 October 1986. On that day, MiG-23 fighters of the 982 IAP flown by Captain Pavlov and Senior Lieutenant Garkavenko, intercepted an Iranian Boeing 747, forcing it to land at Yerevan-Zapadny airport in Soviet Armenia.[24] Travel by air was, and remains, the safest mode of transportation, yet the confrontation between superpowers resulted in some truly tragic exceptions to that general rule.

8
ESCAPE FROM PARADISE

Many who actually had to live there, found the 'workers' and peasants' paradise' not that idyllic, however, leaving the Soviet Union was by no means easy. One possible way to escape the lack of material amenities and freedom, as well as various other problems, was to hijack an aircraft entering it either as a passenger or a pilot.[1] Numerous civilian and military aircraft were thus *commandeered* with the PVO's fighter aviation being no exception, as it was not spared an embarrassing case of defection.

A Non-Event?
Before relating a case which, without doubt took place, a few words need to be devoted to an event which, whilst not impossible, is unlikely to have actually happened. Namely, it has been claimed that in 1961, a disappointed Soviet pilot flew his Su-9 interceptor to Abadan in Iran.[2] Supposedly, the aircraft and its pilot were swiftly secured by officers of the Foreign Technology Division of the US Department of Defence. Reportedly, within just 24 hours, the Su-9 was disassembled and transported to the US with the pilot following shortly thereafter.

However, whilst sounding credible, at first glance and upon closer examination, the story seems to lack merit. Firstly, only very basic information, without any details about this incident, is available, whilst much has been published about other Soviet-made aircraft which, by whatever means, fell into American hands, as well as about defecting pilots.[3][4] In addition, there does not seem to be a solid source to attribute it to.[5] Moreover, as far as the author was able to ascertain, Russian and other post-Soviet sources do not provide confirmation of it ever having happened either. Therefore, in the author's opinion, unless some new and credible information emerges, the story that a Soviet Su-9 fighter-interceptor defected to Iran in 1961 should be dismissed as lacking factual basis. Although speculative, it may well be that the origin of this claimed defection lay in the acquisition of MiG-21s by the Americans in 1960s. This is possible insofar as at that time in the West, due to the lack of available information, the MiG-21 was sometimes confused with similar Soviet fighters. A case in point being that the Jane's All the World's aircraft 1960-1961 edition listed the 'Fishbed' (ASCC/NATO-codename for the MiG-21) as a Sukhoi design and used an illustration of the Su-9.

Defector's Demise
Notably, whilst many defections were successful, not all resulted in the defector making good his escape. In one known instance, PVO fighters put an end to an attempted escape from the Soviet Union with fatal results for the would-be defector.[6] The man in question, Pavel Skrylev, was a pilot first discharged from the Soviet military and subsequently, dismissed from civilian aviation apparently due to his problems with alcohol. Unable to find suitable employment as a pilot and obviously dissatisfied with the way things turned out, he decided to try his luck abroad. With this in mind, he commandeered a civilian An-2 (ASCC/NATO-codename 'Colt') single-engine biplane from Tuapse airfield in Georgia. Having taken off, the defector flew over the Black Sea with the apparent intent to reach Turkey.

Once it was realised that a defection attempt was in progress, a Yak-28P fighter-interceptor of the 171st IAP with Captain Parfilov at the controls, was scrambled and gave chase. Since it quickly became clear that the pilot was the only man aboard the aircraft, an order to shoot it down was issued. The jet-powered Yakovlev caught up with the much slower Antonov without a problem. However, issues started when the former tried to engage the latter. Namely, it proved nearly impossible to lock-on the small biplane flying low over the sea and when finally, the target was acquired, a missile launched against it failed to hit.

Meanwhile, both aircraft were already over international waters but the Soviets would not give the pursuit up. On the contrary, since a missile armed interceptor failed, a cannon armed fighter took its place. A MiG-17 flown by the 171st IAP's deputy commander Lieutenant Colonel Prishchepa, shot the defecting An-2 down with gunfire. The Antonov was destroyed, crashing into the sea together with the escapee pilot who lost his life. Subsequently, a search was conducted but aside some flotsam which was identified as having originated from the aircraft, little was found. In particular, the body of the defector was not recovered.

Moreover, an investigation was conducted which revealed that airfield security was practically nonexistent and that gaining access to an aircraft, as well as starting it, did not present a serious problem for anyone who wished to do so. As a result, various measures were implemented to change this state of affairs. That said, in the years to come, a number of other Soviet civilian aircraft were either commandeered or highjacked. For example, on 23 September 1976, another former Soviet military pilot Valenty Zasimov, defected to Iran in a civilian An-2.[7] This time, the Soviets failed to intercept the escapee. However, the defection attempt failed nonetheless and both the pilot and the aircraft were returned by Iran to the Soviet Union. Once back in the USSR, Zasimov was sentenced to 12 years in a penal colony – a hard landing indeed.

Soviet Secrets Revealed

Arguably, the best-known case of Cold War defection took place on 6 September 1976 when Senior Lieutenant Viktor Belenko, serving with the 530 IAP of the PVO's fighter aviation, flew a MiG-25P fighter-interceptor from Chuguyevka AB, in the Primorskiy Kray of the Soviet Far East, to Hakodate airport, located on the Japanese island Hokkaido.[8]

Numerous reasons prompted Senior Lieutenant Belenko to take this step, including problems in his professional, as well as personal life: he was overdue for promotion to the rank of Captain and his marriage was failing. Obviously, various shortcomings of the Soviet systems, of which there were plenty, as well as discrepancies between official propaganda and observable reality, also played their part. The Soviets and nowadays Russians, suspected that he might have been earlier recruited by US intelligence agencies.[9] However, there is little evidence to support this assertion.

As for Belenko, the man supposedly had a certain manipulative streak but was also outspoken and assertive, at least by Soviet standards.[10] He kept an equal distance to most people yet made sure not to make himself suspect: he befriended the KGB overseer assigned to his unit, did not listen to American radio broadcasts, and forbade his wife to do so.[11]

Perhaps a little surprisingly, the defection itself was not particularly difficult. On the day that fundamentally changed his life, Senior Lieutenant Belenko and several other pilots were performing training sorties. Initially, he followed the assigned flight plan but then, descended rapidly to low-altitude and headed out to sea in the direction of Japan. However, reaching Japan proved to be the easy part, as touching down on Japanese soil turned out to be much more difficult. Since the JASDF failed to intercept Senior Lieutenant Belenko's aircraft, its fighters could not escort the MiG to a suitable landing site, such as the Chitose Air Base. As a result, he had to land at the aforementioned Hakodate airport and if the situation was not dramatic enough, a passenger aircraft was just taking off when the MiG flew in to land.

Fortunately, a collision was averted but the landing was still problematic, for despite deploying the aircraft's drogue chute, the MiG overrun the end of the runway by ca 240m before coming to a halt. Senior Lieutenant Belenko was taken into custody by the Japanese police but soon enough, he found himself in American care and was flown out of Japan to the United States. He was well prepared to be a valuable defector: in addition to all the information he had by virtue of being a PVO fighter pilot in active service, Senior Lieutenant Belenko also brought the MiG-25's flight manual with him. In order to be able to put even more on the table before making good his escape, he also requested permission (which was granted) to access and read various classified materials.

Thus, Senior Lieutenant Belenko was a prized source of intelligence and when extensively questioned, he gave exhaustive answers. Amongst others, he was asked if Soviet pilots would ram enemy aircraft, with his reply being in the affirmative. As will be related in the following chapter, this was indeed the case.

Obviously, the MiG-25 fighter-interceptor Senior Lieutenant Belenko defected with, was also of great value. It was first covered with tarpaulins, then disassembled and moved from Hakodate to Hyakuri military airfield (currently the facility serves as the civilian Ibaraki Airport) where it was analysed in detail. The analysis was a technical intelligence bonanza, causing a western re-evaluation of

Probably the most famous flight of any MiG-25 ever, was that of a disaffected Senior Lieutenant Belenko from the 530th Fighter Aviation Regiment to Hakodate in Japan, on 6 September 1976. Belenko overshot on landing, bringing his jet to stop about 100m behind the runway threshold, as seen here. The jet was eventually returned to the USSR – though only after a throughout inspection by a team of the Foreign Technologies Division of the USAF. (Albert Grandolini Collection)

the MiG-25, which up to that time, was somewhat overrated. Once the Soviets realised what had happened, they demanded the return from Japan, of both the aircraft, as well as the pilot. They claimed that Senior Lieutenant Belenko had made a navigational error and his refusal to return was the result of him being drugged.

Soviet pressure eventually resulted in the return of the MiG-25 in a dismantled condition. However, Viktor Belenko was first granted political asylum in the United States, subsequently received US citizenship and as far as it is known, adopted well to life in America. To say that the whole affair was very embarrassing for the Soviet Union, is an understatement. Since their most capable fighter-interceptor was now compromised, the Soviets modernised the MiG-25 and went to serious work on its eventual replacement which resulted in the development of the MiG-31.

Finally, it should be noted that a similar incident took place, over a decade later, when on 20 May 1989, Captain Aleksandr Zuyev flew a MiG-29 to Turkey.[12] That, however, was a defection of a pilot and aircraft from the VVS and not from the PVO.

Escape to the Soviet Union
As surprising as it may sound, there were not only defections from, but also a few defections to, the Soviet Union. A little-known incident took place on 30 November 1963 when an Iraqi Air Force pilot, Lieutenant Abdel Rahim as-Salim Zohair, defected to the USSR flying through Iranian airspace.[13] The Iraqi aircraft was apparently not intercepted by the Soviets before landing in Baku, in the Soviet Republic of Azerbaijan. Concerning the pilot's motives for the defection, it was his unwillingness to participate in operations against the Iraqi Kurds. There seems to be no solid information regarding the type of aircraft, which was possibly a de Havilland Vampire, nor the subsequent fate of the pilot.

Arguably, the best-known case of a defection to the Soviet Union took place on 11 September 1970 when a Hellenic Air Force pilot Lieutenant Mikhalis Maniatakis, commandeered a C-47 transport from the Kania airbase on the island of Crete. To give some context to his actions, it needs to be pointed out that in those times, Greece was ruled by a military dictatorship. The Greek pilot flew to Crimea and was intercepted about 65km south of Sevastopol by a Su-15 of the 62 IAP, scrambled for that purpose.[14] It was the Su-15s first real operational interception other than one performed for training purposes. Having been escorted by the Soviet fighter for landing at the Belbek AB, the Greek pilot requested and was granted political asylum in the USSR.

Not everybody was blessed with such gracious treatment. On 25 August 1990, Senior Lieutenant Wang Baoyu escaped to the USSR flying his Shenyang J-6 fighter – the Chinese version of the Soviet-made MiG-19 – serial number 20520, from Jiahoe AB in Mudangijang, to Knevichi airfield outside Vladivostok.[15] Notably, the Chinese jet was not intercepted by the Soviets: at that time, the Soviet Union was in the process of improving its relations with the People's Republic of China and would not allow an event like this to spoil the rapprochement. Therefore, a week later, Senior Lieutenant Wang Baoyu, along with his aircraft, was handed over to Chinese authorities. Needless to say, he was treated harshly and a severe punishment was doled out: he was sentenced to death, but subsequently, committed to life in prison.

9
DESTROY THE TARGET AT ALL COST

Events described in the previous chapter clearly demonstrate that many citizens of the USSR would have gladly left the Soviet Union for good, even at considerable personal risk – and that PVO pilots were no exception. That said, it would be a mistake to assume that if push came to shove, they would not have fought with all their strength to the point of self-sacrifice. On the contrary, whenever warranted by circumstances, PVO pilots proved ready to go as far as possible, even to ram their opponents. Whilst this idea might at first glance, seem bizarre and the Soviets saw ramming as 'exceptional', they still rationalised it on a number of grounds.[1] Namely, it was possible that in a combat situation, AAMs could be rendered useless by countermeasures such as decoys and jamming, whilst guns might prove insufficient to deal with a target such as a bomber armed with nuclear weapons. Yet this kind of target had to be destroyed, no matter what and thus, performing a *taran* under such circumstances,[2] not only made sense but might be the only thing left to do. Whilst bombers armed with nuclear ordnance mercifully, did not appear over the USSR, Soviet pilots still performed two *tarans* downing airspace violators in the process.

A Fight to the Death
As described in the Volume 1, the affair about the downing of the U-2 over Sverdlovsk, on 1 May 1960 and few earlier attempts to intercept English Electric Canberra reconnaissance bombers, have seen Soviet attempts to down their opponents by ramming. Most of the attempts were either not pressed home or otherwise, were not particularly successful. It was only in late 1973, when resulting from the fact that the USSR bordered on Imperial Iran, the first *taran* (ramming manoeuvre) since the Second World War was successfully performed.

To give some background to the event, it has to be pointed out that the so-called 'peripheral areas' of the USSR were frequently 'soft spots', offering plentiful opportunities for Western aerial reconnaissance. The Shah Mohammed Reza Pahlavi II of Iran was a staunch US-ally and his armed forces almost completely equipped with US-made equipment. Unsurprisingly, he went as far as to allow the Americans to use the territory of his country for intelligence-gathering activities – provided Iranian personnel was involved and, in the course of such operations, equipped with and trained to use, some of most advanced and sensitive high technologies.

This is how starting in 1968, the CIA, DIA, NSA, USAF, the Iranian SAVAK intelligence agency and the Imperial Iranian Air Force (IIAF) had arranged a joint operation – apparently codenamed 'Dark Gene' – in the course of which, not only one or another of Iranian-owned Aero Commanders, but especially RF-5A fighter-reconnaissance aircraft of the IIAF, were regularly flying clandestine operations inside the USSR. These were sometimes crewed by US and other times, by Iranian and sometimes by combined crews (at least in the case of Aero Commanders: RF-5A was a single-seat jet).

As this operation went on (apparently, without any of the involved Freedom Fighters being caught by the Soviets), the appetite grew and thus the idea was born, to deploy the much more powerful

McDonnell-Douglas RF-4 Phantom II reconnaissance aircraft for clandestine sorties launched from Iran into the USSR. The plot behind this phase of the operation was quite simple: in exchange for the Shah permitting the Americans to use 'his' air bases, the White House, the Pentagon and the US Congress granted permission for Iran to acquire six RF-4s. Of course, delivery of Phantoms to Iran required training of IIAF personnel to maintain and fly them: correspondingly, Iranian pilots could fly RF-4s with US instructors in the rear cockpit and, if any of them 'happened' to violate the Soviet airspace, the affair could always be explained as an 'innocent accident', occurring during a 'routine training mission'.

The Devil was in the detail: the variant ordered by Iran was designated RF-4E and represented an advanced version of the slightly older version operated by the USAF, the RF-4C. Moreover, due to the requirements of the Vietnam War, McDonnell-Douglas took nearly two years to manufacture the jets in question and deliver them to Iran. 'In the meantime', the Americans 'loaned' a handful of USAF's RF-4Cs to the IIAF, to 'serve for training of the Iranian personnel'. The aircraft in question were anything else but 'just training tools': actually, all were equipped with some of most advanced reconnaissance equipment the USAF was capable.[3]

Starting in early 1971, this phase of Operation Dark Gene came forward quite well, and up to two missions were flown per month, all by mixed US-Iranian crews: usually, the pilot was from the IIAF, whilst the systems-operator in the rear cockpit was from the USAF. However, a day came when a US-Persian reconnaissance flight was not only detected but actively challenged, by the Soviet Air Force.

On 28 November 1973, a RF-4C piloted by Major Shokouhnia from the IIAF (known to have flown RF-5As at earlier times), with Colonel John Saunders in the rear seat, was intercepted by a MiG-21SM of the 982nd Fighter Aviation Regiment (home-based at Vaziani AB), piloted by Captain Gennady Eliseev.[4] What happened next was a supersonic chase, in which the Soviet pilot barely managed to keep up with the low and fast-flying Phantom: the RF-4C was faster than all the earlier MiG-21-variants. Eventually, Eliseev fired both of his R-3S'.

According to the US version of events, Colonel Saunders managed to decoy both of these by deploying photo-flares. Actually, these were near useless against the seeker-heads of Soviet air-to-air missiles because they were lighting at the wrong frequency: Major Shokouhnia evaded the missiles by a hard break. However, his vigorous manoeuvre slowed the Phantom down sufficiently for Eliseev to attempt engaging with the internal GSh-23 autocannon of his jet: the weapon jammed. Having no weapons left at his disposal, there appeared nothing the Soviet pilot could do. His superiors were of different opinion: they ordered Eliseev to ram the Phantom.[5]

Before moving on, it needs to be pointed out that while taking evasive action, the Phantom had lost airspeed, allowing the MiG to cut the range. Now, came the most dramatic moment, as Eliseev manoeuvred his MiG-21 to hit the RF-4C: while the ground control advised him to hit the Phantom's fin with his wing, the pilot decided to do so with the nose of his aircraft and from below.

He impacted on the left side, close to the engine nozzles, probably causing serious damage – if not outright severing – the tailfin of the US-made reconnaissance jet. This was perfectly enough: the stricken Phantom turned turtle, leaving its crew without a choice but to activate their ejection seats: both Shokhouhnia and Saunders made a safe parachute descent and were apprehended by the Soviet police, shortly after reaching the ground. Their Phantom hit the ground at such a high speed that it was completely destroyed on impact.

Under interrogation, the Iranian and American steadfastly maintained that they had been on a training flight and strayed over the Soviet Union by pure accident – exactly along the cover story prepared in advance just in case of a situation like this. Since the Soviets could not prove the opposite – because the Phantom was completely destroyed in the crash – they were forced to, grudgingly, accept this version of events.[6]

Eventually, the downed RF-4-crew were saved from a protracted stay in the 'workers' and peasants' paradise' by a fortunate coincidence. It so happened that at about the same time, a cartridge with reconnaissance film from a Soviet reconnaissance satellite landed on the 'wrong' side of the border to Iran: indeed, on one of oilfields outside Abadan. Under the given circumstances, Moscow

Captain Genady Eliseev, in the cockpit of a MiG-21, earlier during his career. (Tom Cooper Collection)

Typical memorial for Eliseev, containing an illustration of his *taran* manoeuvre and a short description with his photos. Such memorials could be found on dozens of PVO and V-VS bases during the Cold War. (Tom Cooper Collection)

was more than happy to exchange the cartridge with precious intelligence photographs for two seemingly 'incompetent' airmen who could not keep their bearings.

Tragically, the story of Captain Genady Eliseev ended entirely differently: he was killed when slamming his MiG against the RF-4. The Soviet success thus came at a high price, for the pilot died the hero's death while destroying his target literally, 'at all cost'. On 14 December 1973, Eliseev was posthumously awarded the title of the Hero of the Soviet Union.[7]

His feat did not receive great publicity but was not kept secret either. For example, the incident was related in the No. 3/1980 issue of the Polish magazine *Przegląd WL i WOPK,* which featured an article by Colonel Kazimierz Stec describing the event.[8] In addition, foreign pilots who were deployed to Astrahan in the 1980s for live firing exercises, recall that Captain Eliseev was much revered by their Soviet hosts. Moreover, Captain Eliseev who was laid to rest in his native Volgograd, was commemorated by a number of monuments and tablets with one of Volgograd's streets also being named after him.

The Argentine Connection

The second instance of Cold War ramming started at Larnaca in Cyprus, from where a Canadair Limited CL-44 transport (c/n 34, reg. LV-JTN) owned by Transporte Aereo Rio Platensean (an aviation transport company registered in Argentina), took off.[9] Behind the controls was Hector Cordero with Jose Burgueno and Hermete Boasso being the other crew members (the trio were Argentine), whilst the fourth man aboard was a British citizen, Stuart McCafferty. The latter was a broker handling military equipment purchases as the aircraft was not just an ordinary cargo plane making a routine transport flight.

From Cyprus, the CL-44 flew to Tel Aviv where it picked up a load of aircraft spares and tyres destined for Iran's air force which needed them in order to sustain its operability in the war with Iraq. As it later transpired, the aircraft had already made two flights to Teheran and on 18 July 1981, was returning from a third but this time it did not make it back. Considering that the Iraq-Iran war was also intensively fought in the air, the Argentinian pilot thought that the best way to avoid being caught up in air combat or shot down by ground-based air defences, would be to fly a northern course, 'hugging' the Soviet border before reaching Turkey. Once in Turkish airspace the plane and its crew could consider themselves safe and would continue flying to Cyprus.

It had worked twice but on the third occasion, it turned out differently as the Argentinian aircraft crossed the Soviet border and was actually flying over Soviet Azerbaijan. Whilst the CL-44 never intruded deep into Soviet airspace, the violation was enough to get Soviet air defences running in full gear. Since the area right along the border was not covered by SAMs, a number of interceptors had to be scrambled in order to deal with the intruder. Among them was a Su-15TM 'bort' red 30 of the 166 IAP flown by Captain Valentin Kulyapin.[10] As it happened, he was the only one to intercept the airspace violator as all his colleagues broke off the chase due to running low on fuel.

Once the foreign aircraft was intercepted, the Soviet pilot took station off its left wing – that is, he positioned himself between the CL-44 and the border. He gave the intruder a closer look, reporting that it was a four-engine aircraft with its windows covered.[11] While flying on a parallel course, Captain Kulyapin could clearly see the

A Su-15TM armed with R-98R (left side) and R-98T (right side) air-to-air missiles, standing QRA in the early 1980s. Ironically, this type became the PVO's interceptor which became involved in most of incidents of the 1970s-1980s that resulted in exchange of fire. (Albert Grandolini Collection)

As a consequence of multiple intercepts during the 1970s and early 1980s, by 1984, all the surviving Su-15TMs of the PVO had their arsenal significantly expanded. Launch rails for R-60 short-range, air-to-air missiles were installed on small hardpoints next to the wing root, whilst under fuselage stations were regularly occupied by two UPK-23-250 gun pods, each of which was packing a twin-barrel GSh-23 autocannon with 250 rounds of ammunition. (Albert Grandolini Collection)

crewmen in the other aircraft's cockpit – who must also have seen him – and tried to signal them but was ignored.

In a highly propagandised account of the event, Soviet sources claimed attempts were also made to contact the intruding aircraft by radio – but to no effect.[12] The foreign aircraft made a turn, forcing the Soviet pilot to manoeuvre out of its way in order to avoid a collision. If going by the Soviet account, this was a deliberate hostile act by the intruder. However, it seems that once the CL-44's pilot realised that he was not only over the Soviet Union but also had unwelcome company in the form of a Soviet interceptor, he decided to leave the Soviet airspace and turned on a course for the border.

This is a much more rational explanation than the idea of a transport trying to physically challenge a fighter over the latter's home turf. Whilst at first glance it may appear that a propeller-driven aircraft had no chance to make an escape when chased by a jet, it should be pointed out that the incident took place right next to the border with the Araks River, with it being actually in sight.

Unfortunately for those on-board, the CL-44 never made it, as the Soviet pilot wanted to get his pray whatever the cost. Captain Kulyapin radioed ground control that the airspace violator not only failed to follow his instruction but also acted in a hostile manner by forcing the Su-15 to get out of its way and was trying to escape across the border. In response, the ground control ordered the intruder to be destroyed. This was, however, easier said than done.

The Soviet interceptor was, by that time, in a seemingly perfect position right behind the CL-44 but in order to get sufficient clearance from the target so as to be able to launch an air-to-air missile, the Su-15 had to drop back a certain distance.[13] This was not an option as the airspace violator was rapidly approaching the border and once he crossed it, the Soviet pilot could not pursuit him, much less open fire. Since time and distance or rather the lack of both, prevented Captain Kulyapin from engaging the intruder in a conventional way, he decided to ram it.

In order to do this, Captain Kulyapin manoeuvred his Su-15 under the CL-44's tail and struck it from below at 14:44hrs Moscow time as indicated by the Soviet fighter's cockpit chronometer. It was a crippling collision for both aircraft: the Soviet pilot was showered by glass as the canopy disintegrated, his mount started to vibrate violently and the control system ceased to function, leaving him with no option other than to eject. Captain Kulyapin passed out for a brief moment as the tremendous forces of an ejection are enough to knock a man out; he regained his senses when he was dangling under the parachute. Being aloft, he had a bird's view of the unfolding events, he witnessed the intruding aircraft – its tail visibly damaged with the right horizontal stabiliser missing – first spiral towards the ground and finally impacting producing a fireball. The Soviet pilot safely floated down to earth by parachute whilst the four men aboard the CL-44 had no means of escape and all perished, buried under a heap of blazing wreckage.

The Soviets quickly arrived at the scene removing the bodies first which were transported to a morgue in Yerevan and subsequently, handed over to Argentine and British authorities.[14] Before the CL-44's wreckage was cleared, Leopoldo Brave who was at that time Argentina's ambassador to the Soviet Union, was allowed to visit the crash site. There was little to see but twisted and charred debris, except that on one piece of wreckage not affected by fire, most likely the tip of the vertical stabiliser, the diplomat was able to spot the Argentine flag. Interestingly, Ambassador Brave was informed by the USSR's Ministry of Foreign Affairs that the Soviet pilot was also killed in the incident.

Very likely, the Soviets wanted to prevent the Argentinian diplomat from even asking about the possibility of speaking to the person who caused the death of his countrymen. At that time, the Soviet only publicly announced that a foreign aircraft was destroyed which collided with a Soviet one, having first violated the country's airspace flying from the direction of Iran.[15] For the time being, nothing more was publicly stated, and the Soviets obviously intended to keep the whole incident as low key as possible.

After a few years their attitude changed and Captain Kulyapin's story was covered in detail by the Soviet military's official newspaper the *Krasnaya Zviezda* (see endnote no 192). The account of the incident was published, together with a large portrait of Kulyapin who was decorated with the Order of the Red Banner, as well as promoted.

Aside the two instances described, there were a few other Cold War incidents when Soviet and Western aircraft collided, for example the case of the Soviet Su-27 and Norwegian 'Orion' related in Chapter 5. However, that incident resulted from aggressive, as well as reckless, behaviour of the airman involved and was not an intentional ramming attack. Closing the subject, it should be pointed out that the wisdom of risking one's life and losing an aircraft in order to bring down an opponent, is highly questionable. Technically reliable weapons employed by well-trained men should be utilised for that purpose. Having said this, there is no doubt that it took considerable personal courage to perform a *taran*. To keep things in perspective, the two instances of ramming were in fact, marginal occurrences considering numerous Cold War shootdowns and an even larger number of interceptions.

10
ON THE LIGHTER SIDE

Light aircraft, sports aircraft and similar, presented a serious problem for Soviet Air Defences. Their small size resulted in weak radar returns, made them hard to spot visually and their thermal signature was also low. In addition, their slow speed meant that an intercepting fighter risked stalling. As already mentioned in Chapter 5, Iranian light aircraft made intrusions into Soviet airspace on a few occasions, falling victim to the PVO.

Incursions into Soviet airspace by such aircraft were not limited to the USSR's southern 'underbelly' but in fact, took place around its periphery. For example, in July 1969, a US airman, George Patterson, violated the Soviet border in the Black Sea region flying a sports plane.[1]

Another incident took place on 25 July 1976 when a Finnish Cessna 150 Aerobat intruded into Soviet airspace. A Su-15 of the 431 IAP was scrambled in order to intercept the airspace violator but the Cessna landed at the Soviet auxiliary airfield at Alakurtti.[2] There, its pilot refuelled the aircraft from a fuel canister he had taken with him for that purpose. After taking off, the Cessna headed further eastwards. Finally, the Sukhoi's pilot sighted the airspace violator through a break in the clouds but was unable to intercept it. Two more Su-15s and a MiG-15UTI (ASCC/NATO-codename 'Midget') were scrambled but failed to intercept the aerial intruder. The Cessna flew on for more than 300km before attempting a forced landing on a clearing in Karelian woods. The aircraft turned over on its back coming to rest with the undercarriage pointing skywards but fortunately, the pilot and the passenger escaped serious injury. Both tried to hide in the forest but were apprehended by residents of the nearby village Vorone, who handed them over to Soviet border troops. The incident described clearly demonstrates practical difficulties when it came to intercepting light aircraft.

Squaring the Red Square

In the evening hours of 28 May 1987, a civilian Cessna landed literally, in the middle of Moscow and a young pilot, who was a German national, emerged from its cockpit. The man in question was Mathias Rust, who had barely reached adulthood being at that time, just 18 years of age.[3] It was a most unusual event by any standard and it was the culmination of a chain of events which begun some two weeks earlier.

On 13 May 1987, Rust who was a licensed amateur pilot with ca 50 hours of flying experience, took off from Uetresen airfield in the vicinity of Hamburg in West Germany (the country being still divided at that point of time). The aircraft he flew was a rented Cessna F172P (reg. D-ECJB) which he modified by removing some of the seats in the cabin and replacing them with auxiliary fuel tanks, thus giving it extra range. His initial destination was the Faroe Islands, next was Iceland from where he first flew to Norway and then to Finland. This aerial tour of northern Europe took about two weeks, but it was only a lengthy overture to the main event.

Around noon on 28 May 1987, Rust refuelled at Helsinki-Malmi airport in Finland, informed flight control that he would be flying to Stockholm in Sweden and once more took to the air. Once airborne, he turned to the east and did not respond to flight control. After he disappeared from radar screens, the Finns presumed an emergency took place and organised a search and rescue operation.

Little did they know that Rust was not only still airborne but was about to fly into Soviet airspace with the intent to reach Moscow and land there. Before moving on, an obvious question arises, namely what Mathias Rust's motives were. The man himself stated that he wanted to be a messenger of peace although in hindsight, admitted that his actions were irresponsible.

Having flown over the Baltic, Rust crossed the Soviet coastline over Estonia (a Soviet Republic at that time) and assumed a heading towards Moscow. He was detected by PVO's radars at 14:29 and since his aircraft failed to respond to IFF queries, it was designated as 'Target 8255'.[4] He flew through the engagement zone of three SAM batteries belonging to the 54th Air Defence Corps but they were not granted permission to open fire. Instead, a MiG-23 of the 656th IAP with Senior Lieutenant Puchnin, was scrambled from Tapa airbase to visually observe the airspace violator. The Soviet pilot spotted Rust's Cessna, but he too, was denied the permission to engage it.

One cannot help but to ask why the PVO failed to take decisive action to halt Rust's flight; the primary reason was, after the tragic shoot-down of the Korean Boeing described in Chapter 7, Soviet Air Defence personnel were reluctant to act, in particular to use physical force, against a civilian aircraft. As things stood, Rust flew deeper and deeper into Soviet territory, being also greatly aided by fortunate coincidences.

It so happened, that in the vicinity of Pskov, numerous aircraft were performing training sorties. Since many young and inexperienced pilots frequently failed to correctly set their IFF transponders, all aircraft flying in the area, including Rust's, were assigned 'friendly' status by military air control.[5] He was again, falsely identified as 'friendly' in the vicinity of Torzhok due to a similar situation with numerous aircraft flying around. In the latter case, intensive aerial activity was due to an air crash having taken place there on the previous day but the result for Rust was the same.

Thus, he continued to fly unchallenged, getting closer and closer to Moscow. Finally, at about 19:00, Rust's Cessna reached the Soviet capital and he looked for a suitable landing site. Initially, he considered landing inside the Kremlin but discarded the idea on the grounds that if he touched down behind the Kremlin's walls, he could be swiftly taken into custody without anyone noticing it. This would obviously deprive him of publicity and make his endeavour pointless. For this reason, he wanted to land on the Red Square but numerous pedestrians walking about made it impossible.

Finally, Rust decided to land on the Bolshoy Moskvoretsky Bridge by the St. Basil's Cathedral where he indeed touched down and from where he taxiied towards the Red Square before coming to a halt. The unfolding events were filmed with a video camera by a British national. Having stopped his aircraft, Rust exited the cockpit, attracting the curiosity of passers-by and then, patiently waited for the authorities to arrive. They appeared some two hours later taking him into custody.

Mathias Rust seen leaning on the rear fuselage of the Cessna F172P following his landing in Moscow, on 28 May 1987. (William Cobb Collection)

A light civilian aircraft landing in the middle of Moscow, ridiculed the Soviet military and the PVO in particular.[6] Therefore, rather unsurprisingly, the incident resulted in nothing short of a purge: the USSR's defence minister Marshal of the Soviet Union, Sergey Sokolov as well as the PVO's commander Chief Marshal of Aviation, Alexander Koldunov, were both dismissed, as were hundreds of other officers. Whilst solid evidence is lacking, it was widely speculated that the USSR's reformist leader, Mikhail Gorbachev, used the incident as an opportunity to get rid of many old officers who were also political hardliners and thus, at odds with his policies.

As for Rust, he was arrested, tried and sentenced to four years in a labour camp. Fortunately for him, in the waning days of the Cold War, he could expect leniency. As a matter-of-fact, Mathias Rust was never shipped off to a labour camp but was released from prison by the beginning of August 1988, returned to Germany and subsequently lived a rather unsettled life. As for the Cessna with which he made this most unusual flight, after several changes of ownership, it finally became the property of the Deutsches Technikmuseum (German Museum of Technology) in Berlin.[7]

Light Planes Scourge

Rust's stunt inspired numerous emulators who would, for the most part, make shallow violations of the USSR's airspace, sometimes even landing on Soviet territory. For example, two foreign light aircraft made an unopposed overflight of Soviet territory in the vicinity of Tbilisi in early June 1987.[8] In May 1988, for three days in a row, a Cessna 152 flown by the Norwegian pilot Andreas Sommers, penetrated about 2–3km deep into Soviet airspace. Despite being scrambled, Soviet fighters were unable to intercept him before he returned to Norwegian airspace. At the other end of the Soviet Union, yet another airspace violation by a light aircraft took place on 29 March 1989, when an American Cessna entered USSR's airspace in the vicinity of Ratmanov Island. On 9 June, the same year, a German pilot, Hans Schneider flying from Turkey, landed on Batumi airfield (Soviet Georgia), left a bouquet of flowers on the runway, and took off. The stunt was performed in a swift manner as the border was relatively close and as a result, Schneider flew away before the Soviets could react.

However, not all such airspace violators managed to get away. On 2 June 1991, Soviet fighters forced to land, a light aircraft which penetrated USSR's airspace from the direction of Sweden.[9] Attempts to intercept light aircraft also resulted in one of the Cold War's last incidents. Namely, a M-20 light aircraft (reg. D-EMON) piloted by a German national, was flying from Helsinki to Moscow on a pre-agreed flight route. However, due to a lack of coordination between Soviet civilian air control and the PVO, the latter assumed it was an airspace violator. As a result, jet fighters were scrambled but failed to intercept the German aircraft. Yet, the Soviets did not give up and sent a pair of Mil Mi-24 (ASCC/NATO-codename 'Hind') attack helicopters which intercepted and forced the M-20 to land at Pulkovo airfield.[10] Fortunately, the incident did not have any serious repercussions and as for the light planes scourge, it ended for good with the end of the Soviet Union.

11
THE CURTAIN FALLS

The first few chapters described the PVO's capabilities and indeed it had many yet, at the same time, the Soviet Air Defence Force also suffered from a number of serious weaknesses. Arguably, the three most significant ones, not necessarily in that order, were that the PVO was obsolete, inflexible, and compromised.

One of PVO's problems was its sheer size: it had so many aircraft, SAMs, radars et cetera that replacing them in short order was impossible. As a result, various aircraft and weapon systems soldiered on long after they had become obsolete. For example, the 472 IAP was equipped with MiG-19 fighters till 1979 when they were officially replaced with MiG-23s. However, since the process to re-equip an entire regiment took time while the unit had to maintain combat readiness, the last MiG-19 were in fact, retained until 1980.[1]

As a side note, the 472nd IAP had the distinction of becoming the final operator of the MiG-23 fighter in the Russian Air Force – retaining the type up to the time when it was disbanded in 1998. Another example was the 641st IAP which had old Yak-28s on charge until 1988, when it finally received new Su-27s.[2] Similarly, other units flew obsolete aircraft and outdated SAMs and radars were also operated, long after their actual combat value had sharply decreased, especially against a foe who would have had the most modern means of air attack at his disposal.

Inflexibility, a serious shortcoming particularly in case of a high intensity war, was another weakness of the Soviet Air Defence Force. This inflexibility manifested itself in several ways. First of all, the PVO had few AWACS type aircraft and as a result, had for the most part, to rely on a network of ground-based radars. The latter was inflexible due to its very nature. Thus, it could be, and indeed was, mapped by US intelligence with potentially devastating consequences should it have come to war.[3] Another manifestation of PVO's inflexibility lay in its training and tactics.

Showing tactical adroitness and adaptability was frowned upon as 'aerial hooliganism' and the career of anyone accused of the latter could suffer gravely. Instead, practising set-piece interceptions was the norm; in addition, many other tactical procedures were rigid and time consuming. That said, there were periodical attempts to reinvigorate training and mock fighter vs fighter air combat was also practised by PVO pilots. However, that did little to change the overall picture and situations requiring swift adaptability, in particularly fighting a tactically demanding enemy, would have more likely than not, proven too much to handle for the PVO.

Tolkachev's Treason

Another crucial weakness of the PVO in the 1980s was that all of its latest weapons and electronics were completely compromised due to treason. A Soviet electronics engineer, Adolf Tolkachev, one of chief designers at the Phazotron design bureau, volunteered to spy for the USA. Between 1979 and 1985 he – intentionally – provided detailed information and technical specifications about various weapon systems as well as aircraft such as: the R-23, R-24, R-33, R-27, and R-60 AAMs, S-200 and S-300 SAMs as well as MiG-29, MiG-31 and Su-27 fighter-interceptors and their radars in particular. How damaging this would have been should the Cold War turned 'hot' is beyond obvious.

The existence of a spy delivering such sensitive information to the Americans was revealed to the Soviets by an ex-CIA officer Edward Lee Howard. This revelation prompted an investigation which led to Tolkachev's arrest: he was taken into custody by KGB's Alfa *Spetsnaz* operatives.[4] Tolkachev was tried, sentenced to death, and executed in 1986. However, before he had been neutralised, he inflicted incalculable damage to the armed forces of the Soviet Union (the PVO in particular), the Eastern Block and Soviet client states.

The end of the Soviet Union also meant the end of the Soviet PVO. The fate of its assets was basically threefold: abandonment, takeover by successor states and last but not least, continued service under the Russian flag. In some cases, PVO bases were just left by its personnel and became something akin to post-apocalyptic theme parks. Abandoned radar sites and also airbases with derelict aircraft, as well as other disused PVO sites can be found throughout the former Soviet Union and toured by anyone curious enough to make the effort.

The takeover of PVO assets by USSR's successor states (other than the Russian Federation) was, more or less, orderly but there were cases when it was forcibly appropriated with some of its aircraft and other equipment, even seeing action during the conflicts which followed the USSR's collapse. Finally, about 65 percent of Soviet PVO assets were taken over by the Russia Federation continuing to form the core of its air defences. However, the obligation to conform with limits imposed by conventional arms reduction treaties and obsolescence, combined with the lack of economic resources for maintenance, meant its continuing reduction throughout the 1990s. The curtain finally fell on the Russian PVO when it was disbanded in 1998 with its remaining assets taken over by the Russian Air Force.[5]

Change of guards: a scene seen on several PVO bases of the mid-1980s, showing a brand-new MiG-31B being towed past a row of old Su-15TMs. At least three regiments that used to fly Sukhois are known to have re-equipped with the new, powerful interceptor, which became known in the West under its ASCC/NATO-codename 'Foxhound'. (Albert Grandolini Collection)

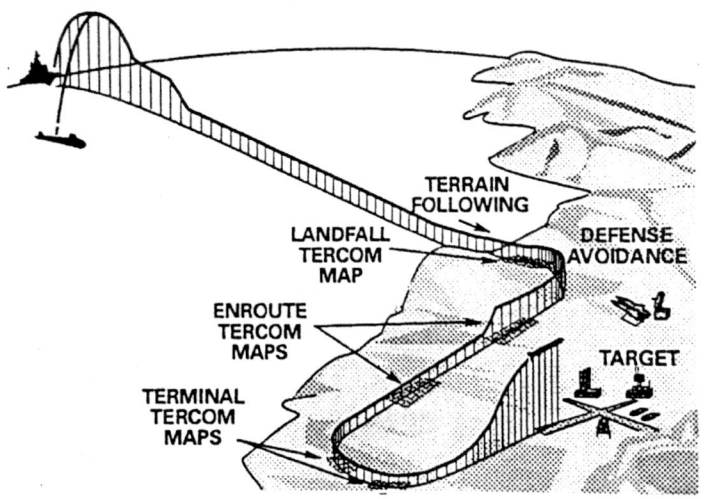

Nowadays, largely forgotten, is the fact that one of the tasks of the R-33-armed MiG-31 was the defence of the USSR from attacks by low-flying cruise missiles – especially RGM-109A and UGM-109As deployed by warships and attack submarines of the US Navy. For this purpose, the jet had to carry a hefty load of R-33 long-range air-to-air missiles and be capable of guiding these precisely with help of its Zaslon radar. (Tom Cooper Collection)

One of very few obvious left-overs from the existence of the PVO still around nowadays, are MiG-31 interceptors, which continue soldiering with the VKS even 30 years after the end of the Cold War and more than 20 years since the end of that type's production. (Albert Grandolini Collection)

APPENDIX I
COMMANDERS OF THE PVO SINCE 1945

1945–1946: General Mikhail Stepanovich Gromadin (1899–1962)

The first commander of the PVO after the end of the Second World War was General Gromadin. He joined the Red Army in 1918 and climbed up through its ranks. He was a veteran of numerous wars and campaigns of the Russian Civil War and commanded air defences in numerous locations during the 'Great Patriotic War', as the Second World War is referred to in the Russian Federation.

Immediately after the victory of May 1945, Gromadin was given the command of the Central Air Defence District. In April 1946, he was appointed the Commander of the PVO once this post was established – albeit, the service was not yet an independent branch of the Soviet armed forces. Gromadin withdrew from this post in June 1948 due to health-related issues and retired in 1954. Between others, he was twice decorated with the Order of Lenin and the Order of the Red Banner.

1946–1952 & 1954–1955: Marshal of the Soviet Union Leonid Aleksandrovich Govorov (1897–1955)

The first commander of the PVO as an independent branch of the Soviet Armed Forces, was Marshal Govorovo. His military career was nothing short of remarkable: it began in 1915 when he was called up to serve with the Russian Imperial Army. During the Russian Civil War, he first served with the 'White' forces, before deserting to the 'Reds'. Fortunate enough to avoid arrest during Stalinist purges of the 1930s, he then climbed up the ranks of the Soviet Army and held numerous commands during the Winter War with Finland, 1939–1940 and during the Great Patriotic War.

Govorov was appointed the Commander of V-PVO between June 1948 and April 1952 when, following numerous failed attempts to intercept one of violators of the Soviet airspace, decision was taken to reorganise the Soviet air defences and improve their effectiveness. Nevertheless, Govorov was given a second tenure (May 1954 – March 1955), with the special task of overseeing a process that was highly valued by the top political and military leadership of the USSR.

Marshal Govorov was awarded with some of highest Soviet military decorations, including the medal and title the Hero of the Soviet Union, Order of Lenin, three times the Order of the Red Banner along with many others.

1952–1953: General Nikolay Nikiforovich Nagorny (1901–1985)

The second commander of the PVO as an independent branch of the Soviet Armed Forces, General Nikolay Nikiforovich Nagorny began his military career during the Russian Civil War, in 1920–1921, when he saw action against anti-Bolshevik insurgency in the Tambov region. Initially specialising in chemical warfare, he was subsequently re-trained in air defence. In this capacity, he took part in the Spanish Civil War. During the Great Patriotic War and immediately after, he held a number of commands, the highest of which was a relatively short stint from June 1952 – June 1953, as the Commander of the V-PVO.

General Nagorny was twice decorated with the Order of Lenin and three times with the Order of the Red Banner.

1953–1954: Air Chief Marshal Konstantin Andreevich Vershinin (1900–1973)

The only man ever to have held the post of both the Commander of the V-VS and the V-PVO, Vershinin joined the Red Army in 1919 and saw extensive action during the Russian Civil War. Initially serving in infantry, he was re-assigned to aviation in 1930, and climbed through the ranks during the Great Patriotic War. He commanded the Soviet Air Force from 1946 until 1949, then the Air Defence Force from June 1953 until May 1954 and then again, the Air Force in 1957–1969 period.

Air Chief Marshal Vershinin was awarded the medal and title of the Hero of the Soviet Union, five times the Order of Lenin, three times the Order of the Red Banner and was decorated numerous other times.

1955–1962: Marshal of the Soviet Union Sergey Semyonovich Biryuzov (1905–1964)

Marshal Biryuzov joined the Red Army in 1922 and rose up its ranks with distinction during the Great Patriotic War. After 1945, he held

General Mikhail Stepanovich Gromadin

Marshal of the Soviet Union Leonid Aleksandrovich Govorov

General Nikolay Nikiforovich Nagorny

Air Chief Marshal Konstantin Andreevich Vershinin

Marshal of the Soviet Union Sergey Semyonovich Biryuzov

Air Marshal Vladimir Aleksandrovich Sudets

a number of high-ranking positions before commanding the PVO from April 1955 until April 1962 – arguably, during the hottest period of the Cold War and the times of its most rapid build-up. Initial failures to shot down one of U-2s underway deep over the USSR, reportedly caused him to exclaim that, if he could, he would turn himself into a missile to bring down the damn intruder.

The highest point in his career came on 1 May 1960, when he was in charge during the famous downing of the U-2 piloted by Francis G Powers. After commanding the PVO, he assumed the command of the Strategic Missile Force and then was appointed the Chief of the General Staff.

Marshal Biryuzov was one of very few high-ranking Soviet officers to meet a violent death in the post-war period. He was killed on 19 October 1964, in the crash of an Ilyushin Il-18 airliner at Mount Avala, near Belgrade (formerly Yugoslavia, nowadays Serbia).

Marshal Biryuzov was awarded the medal and title of the Hero of the Soviet Union, five times decorated with the Order of Lenin, and – between others – three times with the Order of the Red Banner.

1962–1966: Air Marshal Vladimir Aleksandrovich Sudets (1904–1981)

Sudets' military career began in 1925 when he joined the Red Army. Two years later, he qualified as an aircraft mechanic and in 1929, as a pilot. During the 1930s, he held various posts and in 1935, saw action against the Imperial Japanese forces. Sudets served with distinction during the Great Patriotic War and held numerous command posts with the Long-Range Aviation (1955–1962), before being appointed the Commander of the V-PVO, from April 1962 until June 1966.

Air Marshal Sudets was awarded the medal and title of the Hero of the Soviet Union, five times decorated with the Order of Lenin and three times with the Order of the Red Banner.

1966–1978: Marshal of the Soviet Union Pavel Fedorovich Batitskiy (1910–1984)

The longest-serving PVO commander was Marshal Batitskiy. He joined the Red Army in 1924, climbed through its ranks during the period of Stalinist purges and in 1939 was assigned an advisor to the General Chiang Kai-shek in China, before serving in different

Marshal of the Soviet Union Pavel Fedorovich Batitskiy

Chief Air Marshal Aleksandr Ivanovich Koldunov

General of the Army Ivan Moiseevich Tretyak

capacities during the Great Patriotic War. In one of so many ironic twists of fate, in 1950, Batitskiy was appointed the Commander of the Soviet Air Force's detachment in China, where he helped run the final campaign of driving Chiang Kai-shek's forces from the mainland.

Batitskiy's career remained particularly interesting through the next few years,: he took part in the execution of Lavrentiy Beria (infamous chief of the Soviet secret police, deposed in 1953). Batitskiy held a number of high-ranking posts before being appointed the Commander of the PVO for over a decade: from June 1966 until June 1978.

Marshal of the Soviet Union Batitskiy was awarded the medal and title the Hero of the Soviet Union, five times decorated with the Order of Lenin, once with the Order of the Red Banner, and with numerous other medals.

1978–1987: Chief Air Marshal Aleksandr Ivanovich Koldunov (1923–1992)

Koldunov joined the Red Army in 1941, completed his flight training in 1943 and then served as a fighter pilot with distinction, during the Great Patriotic War. Remaining with the service after the victory of May 1945, Koldunov was transferred from the air force to the PVO in 1960 and appointed the Commander of the latter in June 1978. He held that post for nine years, until June 1987

Whilst his career progressed rather smoothly, Koldunov grew rapidly out of touch with modern warfare, as obvious from his 'analysis' of US strikes on Libya in April 1986, which were full of inexplainable errors about aircraft, equipment, weapons, and tactics. The proverbial nail in the coffin of his career came in 1987, when Mathias Rust landed on the Red Square in Moscow: this resulted in Koldunov's dismissal.

Chief Air Marshal Koldunov was twice awarded the medal and title of the Hero of the Soviet Union, three times decorated with the Order of Lenin, and six times with the Order of the Red Banner – in addition to earning himself a large number of other decorations of lower order.

1987–1991: General of the Army Ivan Moiseevich Tretyak (1923–2007)

Tretyak joined the Red Army in 1939, fought with distinction and climbed through the ranks during the Great Patriotic War. He continued his military career afterwards and held a number of high-ranking positions before being appointed the Commander of the PVO in June 1987. However, in 1991, he became involved in the infamous coup, which attempted to preserve the Soviet system but only speeded up its demise: as a consequence, Tretyak was dismissed.

General of the Army, Tretyak was awarded the medal and title of the Hero of the Soviet Union, four times decorated with the Order of Lenin, and three times with the Order of the Red Banner.

General of the Army Viktor Alekseevich Prudnikov (1939–2015)

The last Commander of the PVO began his military career in 1956. He graduated from the Grozny Military Aviation School and climbed up the ladder of the PVO before being appointed to the top position in August 1991. Prudnikov commanded the PVO until the dissolution of the USSR, in December 1991, at which point he transferred to the Armed Forces of the Russian Federation and assumed the command of the Air Defence Force.

General of the Army, Prudnikov was, amongst others, decorated with the Order of the Red Banner.

General of the Army Viktor Alekseevich Prudnikov

APPENDIX II
AERIAL VICTORIES OF THE PVO, 1945–1991

Date	Unit	Aircraft	Pilot	Weapon	Target & Notes
Autumn 1949	no data available	Fighter aircraft	no data available	guns	B-25 reportedly shot down over the Black Sea after dropping parachutists over Ukraine; never confirmed
8 Apr 1950	30th GIAP	La-11	Lt. I. Tezayev	20mm	PB4Y-2 BuNo 59645, 'Turbulent Turtle', VP-26 USN, all 10 killed; first confirmed 'kill' of the Cold War
May 1950	no data available	La-11	Capt. V. S. Yefremov	20mm	P-51D Mustang; never confirmed
26 Dec 1950	523rd IAP	MiG-15	Capt. S. A. Bakhayev Sr. Lt. N. Kotov	23/37mm	RB-29 claimed but never confirmed; possibly RB-29 44-61855, 91st SRS USAF, aircraft damaged, crew survived
6 Nov 1951	88th GIAP	La-11	Sr. Lt. I. Ya. Lukashev Sr. Lt. M. K. Shchukin	20mm	P2V BuNo 124283, VP-6, USN; all 10 killed
13 Jun 1952	11th IAP	MiG-15	Capt. O. P. Fedotov Sr. Lt. I. P. Proskurin	23/37mm	RB-29 44-61810, 91st SRS, USAF; all 12 killed
13 Jun 1952	483rd IAP	MiG-15	Capt. G. Osinskiy	23/37mm	DC-3 79001 'Hugin', 6 Transportflyggruppen, Swedish Air Force (SIGINT mission); all 8 Killed
16 Jun 1952	483rd IAP	MiG-15	Lt. N. Semernikov Lt. I. Yatsenko-Kosenko	23/37mm	PBY-5 47002, Swedish Air Force, SAR for DC-3; crew survived
7 Oct 1952	368th IAP	La-11	Sr. Lt. Zheryakov Sr. Lt. Lesnov	20mm	RB-29 44-61815, 91st SRS, USAF; all 8 killed
18 Nov 1952	781st IAP	MiG-15	Capt. N. M. Belyakov Sr. Lt. B. V. Pushkarev Sr. Lt. A. I. Vandayev Sr. Lt. V. I. Pakhomkin	23/37mm	F9F-5 BuNo 125459, returned safely to USS Oriskany (CVA-34) but damaged beyond economic repair; pilot LTJG Royce Williams claimed 4 MiGs, actually confirmed beyond doubt are 2
16 Feb 1953	368th IAP	La-11	Capt. Lesnov	20mm	F-84 Thunderjet; never confirmed
29 Jul 1953	88th GIAP	MiG-17	Capt. A. D. Rybakov Sr. Lt. Yu. M. Yablonovsky	23/37mm	RB-50 47-145A, 343rd SRS, USAF, 16 killed, 1 survived
1953	no data available	MiG-15 or 17	no data available	23/37mm	RAF Canberra, possibly damaged during a mission to Kapustin Yar
8 May 1954	614th & 619th IAP	MiG-17	Capt. Kurbatov Sr. Lt. Zhiganov Sr. Lt. Kitaychik	23/37mm	RB-47, 91st SRW; damaged by a single cannon shell while conducting the highly successful Murmansk reconnaissance mission
4 Sep 1954	22nd IAP	MiG-17	Capt. N. Serebryakov Capt. P. Kasyanov	23/37mm	P2V BuNo 128357, VP-6, USN, 1 killed, 9 survived
7 Nov 1954	610th IAP	MiG-15	Capt. Kostin Sr. Lt. Serebryakov	23/37mm	RB-29 42-94000, 'Tiger Lil', 6007th Composite Group, USAF, 1 killed, 7 survived

Continued on page 58

Date	Unit	Aircraft	Pilot	Weapon	Target & Notes
3 Dec 1954	535th IAP	MiG-15bis	Capt. P. Byvshev	23/37mm	Tu-14, 49th MTAP, V-VS; 3 killed; fratricide fire
1954	no data available	MiG-17P	Capt. L. I. Savichev	23/37mm	reconnaissance aerostat ('spy blimp')
18 Apr 1955	865th IAP	MiG-15bis	Capt. Rubtsov Capt. Venediktov	23/37mm	RB-47 51-2054, 4th SRS, USAF, all 3 killed
22 Jun 1955	564th IAP	MiG-15	Sr. Lt. Teryayev Lt. Filyutin	23/37mm	P2V BuNo 131515, VP-9, USN; crash-landed and burnt out, no loss of life
27 Jun 1958	976th IAP	MiG-17	Capt. G. F. Svetlichnikov Sr. Lt. B. F. Zakharov	23/37mm	C-118 51-3822, USAF, no loss of life
2 Sep 1958	117th IAP & 168th IAP	MiG-17	Sr. Lt. Kucheryaev Sr. Lt. Ivanov Sr. Lt. Lopatkov Sr. Lt. Gavrilov	23/37mm	C-130 60-528, 7406th Support Squadron, USAF, all 17 killed
16 Nov 1959	no data available	SAM	no data available	SA-75	WS-416L reconnaissance aerostat, shot down near Volgograd (Stalingrad), first Soviet SAM-kill
1950s	179th GIAP	MiG-15 or 17	Capt. Selivanchik	23/37mm	13 reconnaissance aerostats claimed as shot down
1 May 1960	2 Battalion, 57th Anti-Aircraft Missile Brigade	SAM	Maj. Mikhail Voronov	SA-75	U-2A Article 360, pilot F. G. Powers captured (subsequently exchanged)
1 May 1960	4 Battalion, 57th Anti-Aircraft Missile Brigade	SAM	Maj. Shugayev	SA-75	MiG-19S, 365th IAP, pilot killed, fratricide fire
1 Jul 1960	174th GIAP	MiG-19S	Capt. V. Polyakov	30mm	RB-47H 53-4281, 38th SRS, USAF, 4 killed, 2 captured (subsequently exchanged)
4 Aug 1961	no data available	fighter aircraft	no data available	guns	DC-4 EP-ADK, Iran Airways; damaged and forced to make a belly landing, written off, no loss of life
8 Jan 1962	no data available	MiG fighters	no data available	force not used	Caravelle, Sabena Belgian Airlines; forced to land
30 Jun 1962	no data available	SAM	no data available	SA-75 (SA-2)	Tu-104, Aeroflot; all 82 occupants killed; possible fratricide fire (never officially confirmed)
20 Nov 1963	156th IAP	MiG-19S	Capt. Pavlovski	30mm	Aero Commander 500 EP-AEL (c/n 500-846 98), privately-owned; attacked near Mashhad inside Iranian airspace, 3 killed, 1 survived
1964	156th IAP	MiG-19S	Capt. Pechenkin	force not used	Aero Commander 500, forced to land
14 Dec 1965	no data available	SAM	no data available	SA-75	RB-57 63-13287, 7407th Support Squadron, USAF, all 2 killed, never officially confirmed

Date	Unit	Aircraft	Pilot	Weapon	Target & Notes
mid-1960s	2 Battalion, 441st Anti-Aircraft Missile Regiment	SAM	no data available	SA-75	Yak-25M, PVO, crew of two survived; fratricide fire
3 Apr 1966	146th GIAP	Yak-25M	no data available	37mm	reconnaissance aerostat
13 Mar 1967	171st IAP	MiG-17	Lt. Colonel Prishchepa	23/37mm	An-2, 1 killed; shot down during defection attempt
28 Jun 1967	156th IAP	MiG-19	Capt. Stepanov	force not used	L-20 Beaver, Imperial Iranian Ground Forces, forced to land
29 Oct 1967	865th IAP	MiG-17 & Su-9	Capt. Avtushko Maj. Lysyura Capt. Plakhotni	23/37mm R2-US AAM	reconnaissance aerostat
Jun 1968	146th GIAP	Yak-25M	Maj. Volkv Capt. Vakhrushev	37mm	reconnaissance aerostat
1 Jul 1968	308th IAP	MiG-17	Capt. Aleksandrov Capt. Igonin Maj. Yevtushenko Capt. Moroz	force not used	Douglas DC-8 Super 63CF, Seaboard World Airlines, Flight 253A, forced to land
14 Dec 1969	174th GIAP	Yak-28P	Capt. Chernega	K-8 AAM	reconnaissance aerostat
11 Sep 1970	62nd IAP	Su-15	no data available	force not used	C-47, Greek Air Force, forced to land during defection attempt
06 Oct 1970	no data available	MiG-21	no data available	cannon fire	Iranian Aero Commander 500B EP-ALP damaged but managed to reach Iran and land
21 Oct 1970	166th IAP	Su-15	no data available	force not used	U-8 58-3085, USAF, forced to land
early 1970s	393rd GIAP	Su-11	no data available	K-8 AAM	Su-9, PVO; pilot survived; fratricide fire
21 Jun 1973	976th IAP	Su-15	no data available	force not used	Aero Commander 500; forced to land
3 Jul 1973	865th IAP	Su-9	Capt. Blinov	R2-US AAM	reconnaissance aerostat
17 Oct 1973	62nd IAP	Su-15	no data available	R-98 AAM	reconnaissance aerostat (gondola shot-off)
28 Nov 1973	982nd IAP	MiG-21SM	Capt. G. Eliseev	Rammed	RF-4C, IIAF, all 2 survived (exchanged)
Jun 1974	365th AP	Tu-128	Colonel E. I. Kostenko Lt. Colonel Gaidukov other pilots & radar operators	R-4 AAM	6 reconnaissance aerostats – all Soviet – shot down to prevent them from drifting into the PR China's airspace
19 Aug 1975	562nd IAP	MiG-19S	Capt. Sninshikov	30mm	reconnaissance aerostat*
5 Sep 1975	562nd IAP	Yak-28P	Capt. Sninshikov	K-8 AAM	reconnaissance aerostat*
1975-76	no data available	MiG-23M	no data available	AAM	J-7, PLAAF; never officially confirmed
1975-76	no data available	SAM	no data available	Strela-2	J-7, PLAAF; never officially confirmed
24 Aug 1976	no data available	SAM	no data available	S-125	RF-5A, THK, pilot survived
20 Apr 1978	431st IAP	Su-15	Capt. Bosov	R-98 AAM	B707-321 HL7429, Korean Airlines, Flight 802, damaged and forced to make belly landing, 2 passengers killed
21 Jun 1978	152nd IAP	MiG-23M	Capt. Valery Shkinder	R-60 AAM	CH-47C, Imperial Iranian Army Aviation; all 8 killed

Continued on page 60

Date	Unit	Aircraft	Pilot	Weapon	Target & Notes
21 Jun 1978	152nd IAP	MiG-23M	Capt. Valey Shkinder	23mm	CH-47C, Imperial Iranian Army Aviation; damaged and forced to land, no loss of life
23 Dec 1979	976th IAP (?)	Su-15	Capt. Kondakov	force not used	Cessna, Iran, voluntarily landed upon sighting the Soviet fighter
late 1970	518th IAP	Tu-128	Maj. V. Sirotkin Maj. E. Shchetkin	R-4 AAM	reconnaissance aerostat
late 1970s	518th IAP	Tu-128	Maj. E. Shchetkin	R-4 AAM	reconnaissance aerostat
18 Jul 1981	166th IAP	Su-15TM	Capt. V. Kulyapin	Rammed	CL-44 LV-JTN (c/n 34), all 4 killed
1 Sep 1983	777th IAP	Su-15TM	Lt. Colonel G. Osipovich	R-98 AAM	B747-230B HL7442, Korean Airlines, Flight 007, all 259 killed
Apr 1984	82nd IAP	MiG-25PD & MiG-23M	no data available	R-40 AAM 23mm	reconnaissance aerostat
8 Jun 1985	Aktyubinsk Test Centre	MiG-23M	Colonel A. Sokovykh	R-24T AAM	An-26, PVO (flying laboratory), all 8 killed; fratricide fire
19 Jul 1985	350th AP	Tu-128	Maj. N. Savoteev Maj. V. Shirochenko	R-4 AAM	reconnaissance aerostat
29 Oct 1986	982nd IAP	MiG-23M	Capt. Pavlov Sr. Lt. Garkavenko	force not used	B747, Iranian Airlines, forced to land
28 May 1987	656th IAP	MiG-23M	Sr. Lt. Puchnin	force not used	Cessna F172P D-ECJB; piloted by Mathias Rust, intercepted but denied permission to open fire
18 Aug 1989	174th GIAP	MiG-31B	no data available	R-33 AAM	MiG-31B, crew ejected but 1 killed; fratricide fire
13 Sep 1987	941st IAP	Su-27P	Sr. Lt V. Tsymbal	collision	P-3B 602, No. 333 Squadron, Royal Norwegian Air Force; damaged, no casualties
2 Sep 1990	no data available	no data available	no data available	force not used	Iranian military aircraft forced to land
3 Sep 1990	431st IAP	Su-15TM	Capt. I. Zdatchenko	R-60 AAM	unmanned balloon
19 Jan 1991	62nd IAP	Su-15TM	Capt. Ivanov	no data available	unmanned balloon; the last 'kill' of the Cold War
2 Jun 1991	no data available	fighter aircraft	no data available	force not used	light aircraft, forced to land
15 Jul 1991	no data available	Mi-24 helicopters	no data available	force not used	M-20, Germany, forced to land; last incident of the Cold War

*Between 11 August and 14 September 1975, MiG-19s, MiG-21s, Su-15s, Tu-128s and Yak-28s of the PVO used guns, unguided rockets, and AAMs to shoot-down eight reconnaissance aerostats, and shot away the gondolas of two additional ones.

DOCUMENTS

Direktivy Genshtaba VS SSSR ot 31 yanvarya i 30 marta 1967 g
Postanovleniye TSK KPSS i SM SSSR ot 24 yanvarya 1986 g
Postanovleniye TSK KPSS i Soveta ministrov SSSR ot 3 fevralya 1956 «O protivoraketnoy oborone»
Postanovleniye TSK KPSS i Soveta ministrov SSSR № 561–233'
Postanovleniye TSK KPSS i Soveta ministrov SSSR № 161-64'
'Postanovleniye TSK KPSS i Soveta ministrov SSSR № 1160-596'
'Postanovleniye TSK KPSS i SM SSSR ot 8 aprelya 1958 goda'
Postanovleniye SM SSSR №376-119 ot 10 iyunya 1971 g'.

BIBLIOGRAPHY

Angelskiy R. & Korovin V., 'Otechestvennyye upravlyayemyye rakety vozdukh-vozdukh', *Tekhnika i Vooruzheniye*, Vol.9, 2005

Barron J., MiG Pilot: *The Final Escape of Lieutenant Belenko* (New York: McGraw-Hill, 1980)

Belous, V. S., *Shchit Rossii: Sistemy Protivoraketnov Oborny* (Moscow: MGTU, 2009)

Belyakov, R. A. & Marmen, Z., *Samolety MiG, 1939–1995* (Moscow: Aviko Press, 1996)

Cheltsov, B., Zarozhdeniye I Razvitiye PVO Strany, *Voyenno-Istoricheskiy Zhurnal*, Volume 12, 2004

Crickmore, P. F., *Lockheed SR-71 Operations in the Far East* (Oxford: Osprey Publishing, 2008)

Crickmore, P. F., *Lockheed SR-71 Operations in Europe and the Middle East* (Oxford: Osprey Publishing, 2009)

Department of Defense, *Soviet Military Power: An Assessment of the Threat, 1988* (Washington DC: Department of Defense, 1988)

Drozdov, S., V. 'Polety, o kotorykh starayutsya ne vspominat', *Krylya Rodinuy*, Volume 3, 2015

Fedosov, E., A. Aviatsiya PVO Rossii I Nauchno-Tekhnicheskiy Progress (Moscow: Drofa, 2004)

Flintham, V., *Air Wars and Aircraft: A Detailed Record of Air Combat 1945 to the Present* (London: Arms and Armour Press, 1989)

Freundt, L., *Sowjetische Fliegerkräfte Deutschland*, 1945–1994 (Diepholz: Freundt Eigenverlag, 1998)

Ganin, S., 'Sistema–125', Tekhnika i Vooruzheniye, Volumes 8, 9, 10, 2003

Ganin, S., 'Sistema–200', Tekhnika i Vooruzheniye, Volumes 11 and 12, 2003, Volumes 1, 2, 3, 4 and 5, 2004

Ganin, S., & Karpenko, A. V., Zenitnaya Raketnava Sistema S-300, *Nevskiy Bastion Magazine* (special issue), 2001

Gwertzman, B., *U. S. Generals in Soviet after their Plane strays*, The New York Times, 23 October 1970

Hopkins, R. S. III, *Boeing KC-135 Stratotanker: More than a Tanker* (Manchester: Crecy Publishing, 2022)

Hopkins, R. S. III & Habermehl, M., *Boeing B-47 Stratojet: Strategic Air Command's Transitional Bomber* (Manchester: Crecy Publishing, 2016)

Kotlobovskiy, A. & Seidov, I., Goryacheve Nebo 'Kholodnoy Voyny', Parts 1 and 2, *Mir Aviatsii*, Volumes 2 (10), 1995 and 1 (11), 1996

Lofgren, S., *Dödligt Drama över Östersjön, Flygrevyn*, Volume 8, 2010

Kulakov, A. F., *K pyatidesyatiletiyu ordenov Lenina I Krasnoy Zvezdy, GNIIP-10* (Moscow: SiDiPress, 2006)

Kushnerev, V. V., *Sovetskaya Voyennaya Aviatsiya na Kavkaze v Gody Kholodnoy Voyny*, Voyennyy Akademicheskiy Zhurnal, Volume 2 (10), 2016

Manoucherians, L., *Project Dark Gene/IBEX*, Iranian Aviation Review, No. 14

Markovskiy, V. & Perov, K., *Sovetskiye aviatsionnyye Rakety Vozdukh-Vozdukh* (Moscow: Izdatelskiy tsentr Exprint, 2005)

Markovskiy, V. & Prikhodchenko, I., *Istrebitel-Perekhvatchik Su-15* (Moscow, Eksmo, 2015)

Merlin, P. W., *Design and Development of the Blackbird: Challenges and Lessons Learned* (Washington: American Institute of Aeronautics and Astronautics, 2009)

Mikhailov, V. N., *Yademyye Ispytaniya SSSR, Tom 2* (Moscow: Sarov, 1997)

Medved, A. N., *Aviakollektsiya No. 7, 2008: Semeystvo Samoletov Yak–26, Yak–27 i Yak–28* (Moscow: ZAO Redaktsiya Zhurnala Modelist-Konstruktor, 2008)

Moroz, S., *Aviakollektsiya No. 10, 2010: Istrebitel MiG-23* (Moscow: ZAO Redaktsiya Zhurnala Modelist-Konstruktor, 2010).

Moroz, S., Prikhodchenko, I., Kolobanov, V., Istrebitel Su-27 (Moscow: Izdatelskiy Tsentr Eksprint, 2004)

Morozov, V. G., 'Vsevidyashcheye Oko Rossii', *Nezavisimove Voyennove Obrozreniye*, 2000-04-14

Nikolskiy, M., Voyenno-Tekhnicheskaya Seriya No. 95: MiG-31 Strazh Rossiyskogo Neba (Kirov: Kirovskoye Obshchestvo Lyubitelev Voyennoy Tekhniki I Modelizma, 2000)

Pedlow, G. W., Welzenbach, D. E., The Central Intelligence Agency and Overhead Reconnaissance: the U-2 and OXCART Programmes, 1954–1974 (Washington: Central Intelligence Agency, History Staff, 1992)

Peebles, C., *Shadow Flights: America's Secret Airwar against the Soviet Union: A Cold War History* (Presidio Press, 2000)

Peterson, M. L., *Maybe You had to be There: The SIGINT on Thirteen Soviet Shootdowns of US Reconnaissance Aircraft*, Cryptologic Quarterly, Volume 12, No. 2, Summer 1993

Richardson, D., *Techniques and Equipment of Electronic Warfare* (London: Salamander Books Ltd., 1985)

Rigmant, V., *Aviakollektsiya No. 3, 2009: Otechestvennyye Samoloty i Vertoloty DRLO* (ZAO Redaktsiya Zhurnala Modelist-Konstruktor, 2009)

Rigmant, V., *Aviakollektsiya No. 1, 2007: Dalnyi Perekhvatchik Tu-128* (Redaktsiya Zhurnala Modelist-Konstruktor, 2007)

Sahonchik, S. & Semenov, A., *Piloty Zabytoy Voyny* (Moscow: Gorizont, 2021)

Seidov, I., 'Na yuzhnykh rubezhakh Kholodnoy Voyny', *Aviatsiia i Vremia*, 03/2000

Shirokorad, A. B., *Istoriya Aviatsionnogo Vooruzhenya* (Minsk: Kharvest, 1999)

Sobolev, V. A., Lubyanka, 2: *Iz Istorii Otechestvennoy Kontrrazvedki* (Moscow: Mosgosarkhiv, 1999)

Yakubovich, N., *Aviakollektsiya No. 5, 2010: Mnogotselevoy Samolet MiG-25* (ZAO Redaktsiya Zhurnala Modelist-Konstruktor, 2009)

Yakubovich, N., *Aviakollektsiya No. 9, 2011: Istrebiteli-Perekhvatchiki Su-15* (ZAO Redaktsiya Zhurnala Modelist-Konstruktor, 2009)

Yakubovich, N., *MiG-31: Neprevzoydenny Istrebiteli-Perekhvatchik* (Moscow: Eksmo, 2018)

Yakubovich, N., *Aviakollektsiya No. 3, 2019: Istrebiteli-Perekhvatchiki Su-9 i Su-11* (ZAO Redaktsiya Zhurnala Modelist-Konstruktor, 2009)

Zaretskiy, B. L., 'Voyska PVO Strany: Vzlety i Padeniya', *Vozdushno-Kosmicheskaya Oborna*, No. 3/2012

Online Sources

ACIG.org (website of the Air Combat Information Group, online 1999-2017; acig.org/acig.info, still available on the Internet Archive)

Aid M: the Soviet Target: the US Intelligence Community versus the USSR (primarysources.brillonline.com)
Airforce.ru forum (airforce.ru)
Aviation Safety Network (aviation-safety.net)
Aviatsiya v Lokalnykh Konfliktakh (skywar.ru)
Bureau of Aircraft Accidents Archives (baaa-acro.com)
Caswell, J., *Information Paper 4 November 1992, Subject: US Cold War losses from 1946 to 1991* (theblackvault.com)
Cooper, T., Krenzel, A., 'Project Dark Gene and Project Ibex' (spyflight.co.uk)
Degani, A., 'Korean Air Lines Flight 007: Lessons from the Past and Insights for the Future' (ntrs.nasa.gov)
Eastern Order of Battle (easternorbat.com)
Fratini, K., 'Swedish Pilots presented with US Air Medal' (usafe.af.com)
Leone, D., 'SR-71 Crew Members tell the Story of a memorable Encounter with a Soviet MiG-31 Interceptor over the Barents Sea' (theaviationgeekclub.com)
National Security Archive (nsarchive.gwu.edu)
Office of Naval Research (nre.navy.mil) currently unavailable
Oprihory, J. L., 'Saving a Blackbird' (airforcemag.com)
Podmoloda, V., 'Polet bez vozvrata' (svvalush.ru) currently unavailable
Poroskov, N., 'Tayna Aerostata PIROG' (sovsekretno.ru)
Raspletin, A. A., 'History PVO' (historykpvo.narod2.ru)
Rempfer, K., 'Finally declassified: Swedish Pilots awarded US Air Medals for saving SR-71 Spy Plane' (airforcetimes.com)
Roadrunners Internationale (roadrunnersinternationale.com)
Ronald Reagan Presidential Library and Museum (reaganlibrary.gov)
RusArmy.com
Schindler, J. R., 'A Dangerous Business: the US Navy and National Reconnaissance during the Cold War' (nsa.gov)
'SR-71 Blackbird: the Cold War's ultimate Spy Plane' (bbc.com)
Stratopedia (stratocat.com.ar)
'The Cuban Missile Crisis: an Eyewitness Perspective' (jfklibrary.org)
Ugolok Neba (airwar.ru)
Vestnik PVO (pvo.guns.ru)
Zenker, S., 'Swedish Space Corporation, 25 Years: 1972–1997' (zenker.se)

NOTES

Chapter 1
1. For additional details, see *Defending Rodinu*, Volume 1.
2. 'Direktivy Genshtaba VS SSSR ot 31 yanvarya i 30 marta 1967 g'.
3. *Vozdushno-Kosmicheskaya Oborona* Volume 3, 2012.
4. 'Postanovleniye TSK KPSS i SM SSSR ot 24 yanvarya 1986 g'.
5. *Vozdushno-Kosmicheskaya Oborona*, Volume 3, 2007.
6. For additional details, see *Defending Rodinu*, Volume 1.
7. Whilst informal and anecdotal this has been reflected in numerous postings on such platforms as rusarmy.com and forums.airforce.ru.
8. Rigmant, *Otechestvennyye samoloty i vertoloty DRLO*.
9. Postanovleniye TSK KPSS i Soveta ministrov SSSR ot 3 fevralya 1956 «O protivoraketnoy oborone» .
10. *Nezavisimoye Voyennoye Obozreniye*, Vol. 2000-04-14.
11. *Soviet Military Power 1988* & www.structure.mil.ru.
12. Contents of this box were kindly provided by the team of the Eastern Order of Battle website (easternorbat.com).

Chapter 2
1. Treaty Between the United States of America and the Union of Soviet Socialist Republics on the Limitation of Anti-Ballistic Missile Systems, 26 May 1972.
2. For additional details, see *Defending Rodinu*, Volume 1.
3. 'Postanovleniye TSK KPSS i Soveta ministrov SSSR № 561-233'.
4. 'Postanovleniye TSK KPSS i Soveta ministrov SSSR № 161-64'.
5. *Tekhnika i vooruzheniye*, Volume 8, 2003.
6. *Tekhnika i vooruzheniye*, Volume 11, 2003.
7. Ganin & Karpenko, *Zenitnaya Raketnaya Sistema S-300*.
8. *Vozdushno-Kosmicheskaya Oborona*, Vol. 5, 2013.
9. Mihailov *Yadernyye ispytaniya SSSR Tom 2*.
10. 'Postanovleniye TSK KPSS i Soveta ministrov SSSR № 1160-596'.
11. Kulakov, *K pyatidesyatiletiyu ordenov Lenina i Krasnoy Zvezdy GNIIP-10*.
12. Mihailov *Yadernyye ispytaniya SSSR tom 2*.
13. 'Postanovleniye TSK KPSS i SM SSSR ot 8 aprelya 1958 goda'.
14. Belous, *Shchit Rossii*.
15. D.S.T.I. Report No. 261.
16. 'Postanovleniye SM SSSR №376-119 ot 10 iyunya 1971 g'.
17. www.structure.mil.ru.

Chapter 3
1. For additional details, see *Defending Rodinu*, Volume 1.
2. Yakubovich, *Istrebiteli-perekhvatchiki SU-9 i SU-11*.
3. *Ibid*.
4. Yakubovich, *Istrebitel-perekhvatchik SU-15*.
5. Aviatsiya i Vremya, Vol. 1, 2003.
6. According to *Aviakollektsiya*, Vol. 3/2019 & Vol. 9/2011. However, according to *Aviatsiya i Vremya*, Vol.6/1998, the Su-9 was withdrawn from service in 1981, and the last were struck from charge the same year.
7. Moroz. *Istrebitel Su-27*.
8. Belyakov. *Samolety MiG*.
9. Moroz, *Aviakollektsiya Vol.10/2010*.
10. Yakubovich, *Aviakollektsiya Vol.5/2010*.
11. For additional details, see Defending Rodinu, Volume 1.
12. Yakubovich, *Aviakollektsiya Vol.4/2014*.
13. Medved, *Aviakollektsiya Vol.7/2008*.
14. Rigmant *Aviakollektsiya № 1 2007*.
15. Nikolskiy, *Voyenno-tekhnicheskaya seriya № 95: MiG-31*.
16. Unless stated otherwise, based on Markovskiy *Sovetkiye aviatsionnyye rakety 'Vozdukh-vozdukh'*.
17. For additional details, see *Defending Rodinu*, Volume 1.
18. The caveat was the MiG-21's RS-2 compatibility; however, this was no preferred armament option for that type.
19. Shirkorad, *Istoriya aviatsionnogo vooruzheniya*.
20. Named after its inventor, the German engineer Karl Gast.
21. Shirkorad, *Istoriya*.

Chapter 4
1. Pedlow et al, *The Central Intelligence Agency and Overhead reconnaissance: the U-2 and OXCART Programmes*.
2. For additional details, see *Hunt for the U-2*.
3. *The Cuban Missile Crisis* https://www.jfklibrary.org.

4. All information regarding the A-12 is sourced from Pedlow *The Central Intelligence Agency and Overhead reconnaissance: the U-2 and OXCART Programmes.*
5. Merlin, *Design and Development of the Blackbird*.
6. *Ibid.*
7. Pedlow. *The Central Intelligence Agency and Overhead reconnaissance: the U-2 and OXCART Programmes.*
8. *SR-71 Blackbird: The Cold War's ultimate spy plane.*
9. Merlin. *Design and Development of the Blackbird.*
10. Crickmore. Lockheed SR-71 Operations *in the Far East.*
11. *Ibid.*
12. Crickmore *Lockheed SR-71 Operations in Europe and the Middle East.* Notably, that book is offering a more detailed analysis of how a MiG-31 could intercept an SR-71, than any of the Russian sources. The said analysis was prepared for Mr. Crickmore's book by the Ukraine-based Russian aviation writer Valery Romanenko.
13. *Ibid.*
14. Robert S. Hopkins III (veteran RC-135-pilot), interview, 09/2022.
15. *Ibid.* Hopkins dismissed such a possibility.
16. Crickmore, Lockheed SR-71 Operations *in the Far East.*
17. Pedlow. *The Central Intelligence Agency and Overhead reconnaissance: the U-2 and OXCART Programmes.*
18. Podmoloda. *Polet bez vozvrata.*
19. Barron. *MiG Pilot.*
20. *SR-71 Blackbird: The Cold War's ultimate spy plane.*
21. Freundt. *Sowjetische Fliegerkrafte Deutschland.* Notably, whilst the 787 IAP was not a PVO but a VVS unit and one based outside the USSR. The subject matter at hand warrants including a description of its operations.
22. Oprihory. *Saving a Blackbird.*
23. *Ibid.* Other sources describing the incident (vide Bibliography) do not mention a Soviet interception attempt, Swedish pilots escorting the 'Blackbird' were unaware of it and the author did not come across any such in Russian sources available to him. Thus, whilst not entirely impossible, this part of the narrative may be a post factum attempt to give more drama to an already dramatic story.
24. Fratini. 'Swedish pilots presented with US Air Medal'.
25. Nikolskiy, *MiG-31 Strazh rossiyskogo neba.*
26. Crickmore, *Lockheed SR-71 Operations in Europe and the Middle East.* Notably, that book is offering a more detailed analysis of how a MiG-31 could intercept an SR-71, than any of the Russian sources. The said analysis was prepared for Mr. Crickmore's book by the Ukraine-based Russian aviation writer Valery Romanenko.
27. *Ibid.*
28. Yakubovich, *MiG-31.*
29. Unless stated otherwise, content of this box is based on Easternorbat.com & 'Soviet Armed Forces, 1945-1991: Organisation and Order of Battle' (ww2.dk).
30. Additionally, on 1 May 1980, the 40th Fighter Aviation Division was activated in Dolinsk-Sokol. This controlled the 308th Fighter Aviation Regiment (replaced by the 41st Fighter Aviation Regiment in April 1983), the 528th Fighter Aviation Regiment, and the 777th Fighter Aviation Regiment. However, the 40th Fighter Aviation Division was disbanded without replacement on 1 May 1986. The 777th IAP was the subordinated to the 24th Air Defence Division again.

Chapter 5

1. *Aviatsya i Vremja*, Vol.3/2000.
2. According to *Aviatsya i Vremja* Vol.3/2000, these were also MiG-19s. However, according to most of other sources (for example: *Mir Aviatsii* Vol.11/1996), the 156th IAP was still equipped with MiG-17s.
3. *Aviatsya i Vremja*, Vol.3/2000 & Leon Manoucherians, interview, 09/2022.
4. www.aviation-safety.net.
5. *Mir Aviatsii*, Vol.11/1996.
6. *Aviatsya i Vremja*, Vol.03/2000.
7. *Ibid.*
8. Markovskiy. *Istrebitel-perekhvatchik Su-15.* According to Voyennyy Akademicheskiy Zhurnal Vol. 2/2016, this incident took place in 1974 and the aircraft involved were MiG-21s: that, however, appears to be a mistake.
9. *Ibid.* and Manoucherians, interview, 09/2022. Manoucherians has provided detailed information on all four Aero Commanders and Shrike Commanders ever acquired and operated by Iran (whether by civilian/private owners or by the government), leading to the clear conclusion that 'only' two of these were actually shot down by the Soviets.
10. *Aviatsya i Vremja*, Vol.03/20003.
11. *Mir Aviatsii*, Vol.11/1996.
12. *Aviatsya i Vremja*, Vol.03/2000 (notably, the source placed this event in 1981).
13. *Aviatsya i Vremja*, Vol.03/2000.
14. *Mir Aviatsii*, Vol.2/1996.
15. *New York Times*, 23 October 1970.
16. In his report *Information Paper 4 November 1992 LTC Caswell lists the U-8's mission type as recon.*
17. Markovskiy, *Istrebitel-perekhvatchik Su-15.*
18. Details on this incident kindly provided by Arda Mevlutoglu, interview, 04/2022 & Tom Cooper, interview, 04/2022. Additional details from Skywar.ru. Notably, Bogatir concluded a successful career with the THK and is still alive, whilst 1st Lieutenant Sahir Beceren was killed on 26 January 1988 – together with 1st Lieutenant Umit Ortac – when their Lockheed TF-104G crashed near the town of Balikesir.
19. Hopkins, *Boeing B-47 Stratojet.*
20. *Ibid.*
21. Hopkins et all, *Boeing Kc-135 Stratotanke.*
22. *Flygrevyn*, Vol. 8/2010.
23. www.airwar.ru.
24. *Mir Aviatsii*, Vol.11/1996.
25. Sahonchik et all, *Piloty zabytoy voyny*. Notably, although otherwise providing a number of details, surprisingly, the source provided no exact date for this incident.
26. *Aviatsya i Vremja*, Vol.03/2000.
27. *Ibid.*
28. airwar.ru.
29. www.airdisaster.ru.
30. *Aviatsya i Vremja*, Vol.06/1998.
31. *Ibid.*
32. www.testpilot.ru.
33. *Aviatsya i Vremja*, Vol.03/2009.
34. *Caswell, Information Paper 4.*
35. *Ibid.*
36. Peterson, *Maybe you had to be there.*
37. Anonymous sources are usually attributed limited credibility, but this information provided by Robert S. Hopkins III must be given serious consideration.
38. Both Tom Cooper and Robert S. Hopkins pointed at the lack of evidence for any loss of an ERB-47H around that date.

39   *Mir Aviatsii*, Vol.11/1996. Notably, even the source expressed doubts about veracity of such claims.
40   The author cannot avoid the conclusion that whilst much has been published in the former Soviet Union, especially in Russia, in regard to Cold War incidents, there is still a lot of research left to be done. In particular, a number of incidents still need to be confirmed or disproven. Additionally, the quality of the information presented varies considerably, many events are related in an anecdotal, almost unserious fashion. For example, a particularly annoying habit of many Russian authors is the omission of an exact date. Having said that, if the amount of information published thus far is anything to go by, it is not beyond the realms of possibility that many of the still existing gaps in our knowledge about Cold War incidents, will be eventually filled.

Chapter 6

1   *Mir Aviatsii*, Vol.11/1996.
2   Krasnaya Zvezda. 19 June 1993.
3   *Aviatsiya i Kosmonavtika*, Vol.6/2007.
4   *Aviatsiya i Kosmonavtika*, Vol. 10/2011.
5   *Aviatsiya i Kosmonavtika*, Vol. 10/2011.
6   *Voyennyy Akademicheskiy Zhurnal*, Vol. 10/2016 .
7   *Mir Aviatsii*, Vol.17/1998.
8   *Aviatsya i Vremja*, Vol.2/1997.
9   Markovskiy. *Istrebitel-perekhvatchik Su-15*.
10 www.stratocat.com.ar.
11 *Mir Aviatsii*, Vol. 11/1996.
12 Poroskov, *Tayna aerostata PIROG*.
13 Zenker, *Swedish Space Corporation 25 years* www.zenker.se.
14 www.8oapvo.net.
15 Known cases will be listed in Appendix II.

Chapter 7

1   *Mir Aviatsii*, Vol.12/1996.
2   airwar.ru.
3   *Mir Aviatsii*, Vol.11/1996.
4   *Central Intelligence Bulletin*, 2 July 1968.
5   After landing at Misawa AB in Japan, about an hour later, Captain Tosolini retracted his apology.
6   Interception and downing of the aircraft according to Markovskiy *Istrebitel-perekhvatchik Su-15*.
7   When the Commander of the PVO, Marshal of the Soviet Union Pavel Batitsky, learned of the situation he ordered the airliner to be forced to land but not to shoot it down. However, by that time the damage has already been done.
8   https://aviation-safety.net.
9   Soobshcheniye TASS ot 30 aprelya 1978 goda.
10 https://aviation-safety.net.
11 All details of KAL007's flight acc. to https://aviation-safety.net.
12 The Soviet perspective of how the events unfolded according to Markovskiy, *Istrebitel-perekhvatchik Su-15*.
13 All details of KAL007's demise acc. to https://aviation-safety.net.
14 'Interview with Colonel Osipovich', *The New York Times*, 9 December 1996'.
15 kal-007-nsa-history.pdf available at National Security Archive https://nsarchive.gwu.edu.
16 'Address to the Nation on the Soviet Attack on a Korean Civilian Airliner', reaganlibrary.gov.
17 Soobshcheniye TASS ot 2 sentyabrya 1983 goda.
18 Soobshcheniye TASS ot 7 sentyabrya 1983 goda.
19 gazeta.ru.
20 *Izvestia*, 11 December 2003.
21 ICAO Press release PIO 1/93 Jan 1993.
22 NASA, Korean *Air Lines Flight 007*, Technical Reports Server (ntrs.nasa.gov).
23 www.regeringen.se.
24 *Voyennyy Akademicheskiy Zhurnal* Vol. 2/2016.

Chapter 8

1   Not a few defections were caused by the fact that the career, the personal life, or both of an escapee was not going in the direction he imagined it should. which is frequently more or less directly implied when their stories are related.
2   Skaarup, *Canadian MiG flights*.
3   For example, a detailed article by John Lowery on American MiG-21s in the June 2010 issue of the *Air Force Magazine*.
4   Barron, *MiG Pilot* is a good but not the only example of defecting Soviet pilot's stories being related in detail.
5   For example, Maj Harold Skaarup in his book *Canadian MiG flights*. when mentioning this incident. cites Wikipedia as his source, although neither the same, nor any other publication is providing reference to any kind of declassified documents or other credible sources. The claim that a Su-9 defected in 1961 to Iran seems to be copied from one publication (online or traditional) to another apparently. without anybody making a serious effort to check if it has any merit to begin with.
6   *Mir Aviatsii*, Vol.37/2005.
7   skywar.ru.
8   Unless stated otherwise, based on Barron, *MiG Pilot*.
9   Podmoloda, *Polet bez vozvrata*.
10 Ibid.
11 Ibid.
12 airwar.ru.
13 skywar.ru.
14 *Aviatsya i vremja*, Vol.01/2003.
15 *Krylya Rodiny*. 3 2015.

Chapter 9

1   *Taktika v Boyevikh Primerakh*.
2   In Russian. as well as other Slavic languages (such as the author's native Polish). *taran* means 'ramming'.
3   Tom Cooper, 'Project Dark Gene' (splyflight.co.uk) & Leon Manoucherians, *Iranian Aviation Review*, Volume 14.
4   According to Easternorbat.com, the 982nd IAP was a V-VS-unit, but due to Vaziani's relative proximity to the border, aircraft based there have maintained QRA to deal with possible airspace violations.
5   'Interview with General Aleksandr Ostanin', *Trud*, 11 December 2003. As of November 1973, Ostanin was the CO 15th Air Defence Corps of the Baku Red Banner Air Defence District. The ground crew responsible for the failure of Eliseev's GSh-23 received a prison sentence as punishment.
6   That the Shokouhnia and Saunder's RF-4 was completely destroyed, and the two have managed to convince their interrogators of their fake story, is proven by the fact that for decades after (in few cases: until this very day), the Soviets were convinced the aircraft they flew. was a Lockheed T-33A Shooting Star, two-seat conversion trainer.
7   warheroes.ru.
8   *Przegląd WL i WOPK* means the WL i WOPK review with the acronym standing for *Wojska Lotnicze i Wojska Obrony Powietrznej Kraju* meaning Air and Air Defence Troops. Interestingly. the article was widely viewed as very controversial and received a number of hostile comments from other Polish officers who

thought the author unwisely wanted to popularise ramming as a tactic of aerial combat.
9. aviation-safety.net.
10. Markovskiy. *Istrebitel-perekhvatchik Su-15*.
11. Indeed, there were none in the fuselage for there was no need for them, since it was not a passenger aircraft but a transport.
12. *Krasnaya Zviezda*, 6 April 1986.
13. According to Markovskiy (in *Istrebitel-perekhvatchik Su-15*) the Sukhoi was armed with two R-98 and the same number of R-60 AAMs, but did not carry gun pods.
14. Whilst the incident took place over Azerbaijan the nearest city to the crash site was Yerevan in Armenia.
15. Soobshcheniye TASS ot 18 iyunya 1981 goda.

Chapter 10
1. *Mir Aviatsii*, Vol.11/1996.
2. Markovskiy, *Istrebitel-perekhvatchik Su-15*.
3. Rust's story, motives et cetera was first related by the German magazine *Stern* to which his family had sold the exclusive rights and then, was subsequently written about by other German newspapers and magazines such as the *Frankfurter Allgemeine Zeitung*, *Der Spiegel* and others, as well as English language press, for example the British *Guardian*. Unless noted otherwise, details of Rust's feat were gathered from such openly available sources.
4. *Nezavisimoye Voyennoye Obozreniye*, 06 April 2001.
5. *Ibid*.
6. 28 May was also Border Guards' Day which made the whole thing even more comical for an observer and embarrassing for the Soviets officialdom.
7. technikmuseum.berlin.
8. *Mir Aviatsii*, Vol.12/1996.
9. *Ibid*.
10. skywar.ru.

Chapter 11
1. *Mir Aviatsii*, Vol.16/1998.
2. *Aviatsiya i Kosmonavtika*, Vol.4/2016.
3. Aid, *The Soviet target*.
4. Sobolev, *Lubyanka 2*.
5. structure.mil.ru.

ABOUT THE AUTHOR

Krzysztof Dabrowski from Poland has a lifelong interest in the subject of military aviation and has written dozens of articles on a variety of related subjects for printed magazines and the ACIG.org/ACIG.info, AeroHisto and The Boresight websites. His particular area of interest is air warfare during the Cold War, the aircraft involved, and the experiences of their crews. He has published a number of books on related topics.